LAW, PRACTICE AND POLITICS OF FORENSIC DNA PROFILING

This collection reviews developments in DNA profiling across jurisdictions with a focus on scientific and technological developments as well as their political, ethical, and socio-legal aspects. Written by leading scholars in the fields of social studies of forensic science, science and technology studies and socio-legal studies, the book provides state-of-the-art analyses of forensic DNA practices in a diverse range of jurisdictions, new and emerging forensic genetics technologies and issues of legitimacy.

The work articulates the various forms of technolegal politics involved in the everyday, standardised and emerging practices of forensic genetics and engages with the most recent scholarly and policy literature. In analyses of empirical cases, and by taking into account the most recent technolegal developments, the book explores what it means to live in a world that is increasingly governed through anticipatory crime control and its related risk management and bio-surveillance mechanisms, which intervene with and produce political and legal subjectivities through human bodies in their DNA.

This volume is an invaluable resource for those working in the areas of social studies of forensic science, science and technology studies, socio-legal studies, sociology, anthropology, ethics, law, politics and international relations.

Dr Victor Toom is scientific staff at the Netherlands Scientific Council for Government Policy, The Hague, the Netherlands.

Dr Matthias Wienroth, Centre for Crime and Policing, Department of Social Sciences, University of Northumbria, Newcastle upon Tyne, UK.

Amade M'charek is Professor of Anthropology of Science at the Department of Anthropology, University of Amsterdam, the Netherlands.

LAW, PRACTICE AND POLITICS OF FORENSIC DNA PROFILING

Forensic Genetics and their Technolegal Worlds

Edited by Victor Toom, Matthias Wienroth and Amade M'charek

Routledge
Taylor & Francis Group
LONDON AND NEW YORK

First published 2023
by Routledge
4 Park Square, Milton Park, Abingdon, Oxon OX14 4RN

and by Routledge
605 Third Avenue, New York, NY 10158

Routledge is an imprint of the Taylor & Francis Group, an informa business

British Library Cataloguing-in-Publication Data
A catalogue record for this book is available from the British Library

ISBN: 978-0-367-33849-7 (hbk)
ISBN: 978-1-032-38528-0 (pbk)
ISBN: 978-0-429-32235-8 (ebk)

DOI: 10.4324/9780429322358

Typeset in Bembo
by Apex CoVantage, LLC

CONTENTS

ACKNOWLEDGEMENTS

This book was largely written during the COVID-19 pandemic 2020–2022. As a result, it is a testament to its contributors' resilience during times of restrictions to meeting family, friends and colleagues; travel bans; working from home; online meetings; teaching students via video link; and perhaps also schooling their own children at home. Add to that mix quarantine and isolation due to 'testing positive', being ill with COVID, and perhaps only slowly growing used to living with 'long COVID'. During these difficult two years, life ran at a different pace than before, with new routines, duties and challenges. Nevertheless, our authors composed and edited their chapters, and you hold the result in your hands.

As editors, we are grateful to those who made this volume possible. First and foremost, of course, we thank all contributors for their chapters and the good humour with which they engaged with our suggestions. Second, some of our colleagues set an extraordinary scholarly example that inspired us. While not everyone was on board to contribute to this volume, we hope that you notice that we follow a path laid out by you. In particular, we would like to acknowledge Simon Cole, Mike Lynch, Helena Machado and Robin Williams for your work, inspiration, collegiality and mentorship. We also want to acknowledge the European Research Council for having provided us with space to meet and work together in Amsterdam and elsewhere through an ERC Consolidator Grant (FP7–617451-RaceFaceID-Race Matter: On the Absent Presence of Race in Forensic Identification). Lastly, very important persons have worked in the background, allowing the book to be made: our partners who have provided support and hot tea during cold nights and the marvellous editorial staff at Routledge.

Victor Toom, Matthias Wienroth and Amade M'charek

BIBLIOGRAPHICAL DETAILS

Editors and contributing authors

Judith Amankwa Addo

School of Law, Northumbria University, Newcastle upon Tyne, United Kingdom
Judith Amankwa Addo (PhD Candidate) is currently pursuing her PhD in Law at Northumbria University, researching young people and criminal appeals in England and Wales. She holds a MA in Criminology and Criminal Justice from Keele University and a BA (Hons) Sociology from Kwame Nkrumah University of Science and Technology – Ghana. Her research interests include youth justice, miscarriage of justice, appeals and forensic science.

Aaron Opoku Amankwaa

Department of Applied Sciences, Northumbria University, Newcastle upon Tyne, United Kingdom
Aaron Opoku Amankwaa (PhD) is an Assistant Professor in Forensic Science, Department of Applied Sciences at Northumbria University. His primary research is focused on three key themes: 1) the evaluation of the effectiveness of forensic science applications and technologies; 2) development of forensic genetic testing and 3) development and evaluation of crime reduction initiatives.

Jaime Enrique Castro Bermúdez

Fiscalía General de la Nación, Colombia, Bogotá
Jaime is an anthropologist, specialist in human rights (ESAP) and master in criminology and victimology (ESPOL). He works as forensic anthropologist expert and judicial investigator of the Technical Investigation Body of the Attorney General's Office in Colombia. He has more than 23 years of experience working in the search for disappeared persons and the recovery and identification of bodies

without identity, in addition to the investigation of violation of human rights cases in the context of the Colombian armed conflict. He teaches forensic anthropology at the Javeriana University in Bogota. Currently, Jaime is working on a master's of anthropology in bioarcheology at the National University.

Ilpo Helén

Department of Social Sciences, University of Eastern Finland, Joensuu, Finland
Ilpo Helén is a professor of sociology. He has written widely on biopolitics and the political economy of contemporary genetics and other biomedical technologies. Among his publications are 'Making up families: How DNA analysis does/does not verify relatedness in family reunification in Finland' (with Anna-Maria Tapaninen) (*BioSocieties* 15, 376–393) and 'Biological citizenship across the borders: Politics of DNA profiling for family reunification' (*Distinktion: The Scandinavian Journal of Social Theory* 15, 343–360).

Roos Hopman

Humanities of Nature, Museum für Naturkunde/Institut für Europäische Ethnologie, Humboldt-Universität zu Berlin, Germany
Dr Hopman conducted her PhD research in light of the ERC funded RaceFaceID project (headed by Professor Dr Amade M'charek), where she focused on the face-making technology of forensic DNA phenotyping: techniques aimed at predicting physical characteristics from biological traces found at crime scenes. Building on ethnographic fieldwork in genetic laboratories, she demonstrated how race became differently enacted in forensic DNA phenotyping practices. Currently, she is a postdoctoral researcher working between Humboldt University and the Museum für Naturkunde in Berlin, in a Berlin University Alliance (BUA)– funded project. Her focus in this project is on biodiversity-centred data collection and digitisation practices.

Christopher James Lawless

Department of Sociology, Durham University, Durham, United Kingdom
Dr Christopher James Lawless is an associate professor in the Department of Sociology, Durham University. His primary research interest concerns the sociological significance of forensic and biometric technologies. He has published outputs on forensic and biometric policymaking, the social construction of law-science relationships and the ethical impact of forensic technology. He has assisted with a number of UK and devolved parliamentary inquiries relating to forensic policy, biometric legislation and facial recognition. Dr Lawless also pursues more general interests in policymaking and planning concerning social and technological infrastructures. His work draws upon and seeks to develop various conceptual resources associated with science and technology studies.

Luiza Louzada

Human Rights Clinic of the Federal University of Paraná (CDH-UFPR), Curitiba, Paraná, Brazil
Luiza is a lawyer and researcher. She holds a master's degree in social and legal sciences from the Fluminense Federal University (PPGSD/UFF) and a PhD from the

Institute of Social Medicine of the Rio de Janeiro State University (IMS/UERJ). She works in the areas of digital law, human rights, bio-law and bioethics. She was a member of the Forensic Genetics Policy Initiative Steering Committee on the preparation of a best practice report for forensic DNA databases (2017). She is an associate researcher at the Human Rights Clinic of the Federal University of Paraná (CDH-UFPR). Currently, she works as a lawyer in data privacy at an insurance company and is dedicated to the study of the social, legal and bioethical implications of forensic genetics in Brazil.

Amade M'charek
Department of Anthropology, University of Amsterdam, Amsterdam, the Netherlands
Dr. Amade M'charek is Professor of Anthropology of Science at the Department of Anthropology, University of Amsterdam, where she acts as the director of the research group Health, Care and the Body. M'charek is PI of the RaceFaceID project, an ERC-consolidator project on forensic identification and the making of face and race, and co-PI of the NWA project Pressing Matter: Ownership, Value and the Question of Colonial Heritage in Museums. Her work has centred on the (ir)relevance of race in science and society and the material semiotics of identities. Through her recent research on migrant death, she has developed an interest in forensic methods for studying (post)colonial relations, circulations and extractions, which has translated in her current interdisciplinary project, Vital Elements and Postcolonial Moves: Forensics as the Art of Paying Attention in a Mediterranean Harbour Town for which she was awarded an ERC advanced grant.

Carole McCartney
Leicester Law School, University of Leicester, United Kingdom
Dr McCartney is Professor of Law and Criminal Justice. Previously at Northumbria University, where she established the Science and Justice Research Interest Group, and the University of Leeds and Bond University, Australia. Carole has been researching issues around criminal evidence and forensic science for twenty years, and has written on miscarriages of justice, international policing cooperation, DNA and biometrics, forensic science and criminal justice more widely. She currently teaches and researches in the areas of criminal law, criminal evidence and forensic science.

Erin Murphy
New York University School of Law, USA
Erin Murphy is the Norman Dorsen Professor of Civil Liberties at New York University School of Law. Murphy's research focuses on technology in the criminal justice system, with a particular emphasis on forensic evidence. She is an internationally recognised expert in forensic DNA typing and the author of *Inside the Cell: The Dark Side of Forensic DNA* (Nation Books, 2015). She has published numerous other scholarly works in both academic and popular presses and is co-editor of the renowned modern scientific evidence treatise. Prior to joining the NYU faculty,

Murphy was on the faculty of UC Berkeley School of Law, a staff attorney at the Public Defender Service for the District of Columbia and a law clerk to the Honourable Merrick B. Garland.

María Fernanda Olarte-Sierra

Department of Social and Cultural Anthropology, University of Vienna, Austria

Dr. María Fernana Olarte-Sierra is a medical anthropologist and an anthropologist of science with a focus on ethnographic research. She addresses interactions of health, technology, and the body in highly biomedicalized and technological contexts. Recently, she finished a Marie Skłodowska-Curie fellow grant at the University of Amsterdam. In a context of transitional justice, she examined the politics of Colombian forensic scientists' knowledge production practices that co-shape the collective accounting of violence. These practices contribute to judicial and humanitarian actions of victim reparation and justice. Currently, she is a Postdoctoral Researcher at the University of Vienna, in the Institute for Cultural and Social Anthropology, at the group of Medical Anthropology and Global Health where she is working on practices of collective care and healing in war-ridden contexts in Latin America.

Irene van Oorschot

Erasmus University College (ESSB), Erasmus University Rotterdam, the Netherlands

Irene van Oorschot received her PhD (cum laude) from Erasmus University Rotterdam for her ethnographic study of sociological and legal modes of knowing and valuing, which was recently published as a monograph, *The Law Multiple: Judgment and Knowledge in Practice*, by Cambridge University Press. Since then, she has worked collaboratively with the members of the RaceFaceID project team (ERC consolidator grant, PI: Prof. Amade M'charek). Currently, she is a Marie Skłodowska-Curie Fellow at the Life Science and Society Lab in Leuven, Belgium. There, she studies environmental knowledges and valuation practices.

Barbara Prainsack

Centre for the Study of Contemporary Solidarity (CeSCoS), Department of Political Science, University of Vienna, Austria

Neues Institutsgebäude (NIG), Vienna, Austria

Barbara Prainsack is a professor and head of department at the Department of Political Science at the University Vienna, where she also directs the Centre for the Study of Contemporary Solidarity (CeSCoS) and the interdisciplinary research platform Governance of Digital Practices. Her work explores the social, ethical and regulatory dimensions of genetic and data-driven practices and technologies in biomedicine and forensics. Her latest books are *Personalized Medicine: Empowered Patients in the 21st Century?* (2017) and *The Pandemic Within: Policy Making for a Better World* (with H. Wagenaar, 2021).

Vitor Simonis Richter
Federal University of Rio Grande do Sul, Brazil, Porto Alegre
Vitor holds a master's and doctoral degree in social anthropology from the Federal University of Rio Grande do Sul (UFRGS). His main research interest is identification technologies, connecting areas such as anthropology of science, technology, state and law. Currently, he is a Capes/PNPD anthropology post-doc fellow at the Federal University of Rio Grande do Sul and associate researcher at the Human Rights Clinic of the Federal University of Paraná (CDH-UFPR).

Gabrielle Samuel
Department of Global Health and Social Medicine, King's College London, United Kingdom
Gabrielle Samuel is a Wellcome Research Fellow in the Department of Global Health and Social Medicine at King's College London, UK, and Senior Research Fellow at the Wellcome Centre for Human Genetics, Oxford University, UK. Her work explores the social, ethical and regulatory issues pertaining to new and emerging data-driven technologies, with a specific focus on AI and genomics. Her recent area of interest explores the negative environmental impacts of such technologies. Her latest work includes 'Sustainable biobanks: a case study for a green global bioethics', published with colleagues in *Global Bioethics*

David Skinner
Faculty of Arts, Humanities and Social Sciences, Anglia Ruskin University, Cambridge, United Kingdom
David Skinner is associate professor of sociology at Anglia Ruskin University, Cambridge. He has a longstanding interest in the politics of race and science. Much of his work has focused on policing and forensic genetics. Recent publications include 'Race, racism and identification in the era of technosecurity', *Science as Culture*, (2020); 'Forensic genetics and the prediction of race: What is the problem?', *BioSocieties* (2020); and (with Matthias Wienroth) 'Was this an ending? The destruction of samples and deletion of records from the UK Police National DNA Database', *BJHS Themes* (2019).

Noah Tamarkin
Departments of Anthropology and Science and Technology Studies, Cornell University, Ithaca, New York, US
Wits Institute for Social and Economic Research (WISER), University of the Witwatersrand, South Africa
Noah Tamarkin is an assistant professor of anthropology and science and technology studies at Cornell University and a research fellow at the Wits Institute for Social and Economic Research (WISER) at University of the Witwatersrand in Johannesburg, South Africa. He is the author of *Genetic Afterlives: Black Jewish Indigeneity in South Africa*, published in 2020 by Duke University Press. He is currently

writing a book on the introduction and implementation of South Africa's national forensic DNA database, the forensic genetics networks that it has fostered and its implications for postapartheid South African and global politics of surveillance statecraft, human rights and carcerality.

Anna-Maria Tapaninen

Faculty of Social Sciences/Sociology, University of Helsinki, Finland

Anna-Maria Tapaninen is adjunct professor in Social and Cultural Anthropology at the University of Helsinki. Since 2010, she has studied immigration, first by focusing on the use of biotechnologies (DNA analysis and age assessment) in two projects led by Ilpo Helén and funded by the Academy of Finland and the Kone Foundation, respectively. Among her publications is *Making-up families: How DNA Analysis Does/Does not Verify relatedness in family reunification in Finland* (in Biosocieties), together with Ilpo Helén. Her most recent project was Struggles over Home and Citizenship: Neighbourhood Solidarity as a Response to the Refugee 'Crisis', funded by the Kone Foundation. A jointly written book *Travelling Crises: Ethnographic Perspectives on Borders, Homes, and Refugee Journeys* is forthcoming in 2023.

Victor Toom

Netherlands Scientific Council for Government Policy, The Hague, the Netherlands

Dr Victor Toom is scientific staff at the Netherlands Scientific Council for Government Policy. Prior to that, he was a scientific secretary to the Health Council of the Netherlands and a research fellow in the sociology department at Goethe University in Frankfurt am Main, Germany (2016–2019) and at Northumbria University Centre for Forensic Science in Newcastle upon Tyne, England (2010–2016). Enabled by research grants from the European Union, Leverhulme Trust and the Nuffield Foundation, he researched, inter alia, forensic science practices in Bosnia and Herzegovina, New York City, the European Union and Singapore. His articles have appeared in a number of journals, including *Science, Technology, and Human Values*, the *British Journal of Criminology* and *Forensic Science International: Genetics.*

Matthias Wienroth

Centre for Crime and Policing, Department of Social Sciences, University of Northumbria, Newcastle upon Tyne, United Kingdom

Dr Matthias Wienroth is Vice Chancellor's is Senior Fellow at the Northumbria University Centre for Crime and Policing (2020–2023). He studies science and technology as social practices, attending to questions of how they are imagined to contribute to a 'good society'. Applying sociological and ethical analysis and engaging with scientists, practitioners, civil society organisations and policy-makers, he researches the ways that scientific knowledge and emerging technologies shape the ways of (un)knowing human beings and social orders and the implications for technology use and governance. His focus is on life sciences technologies in health, security and justice. He currently works on modes of values/valuations and validations of biometric technologies in forensic and surveillance uses.

Dana Wilson-Kovacs

Department of Sociology, Philosophy and Anthropology, University of Exeter, United Kingdom

Dana Wilson-Kovacs is Associate Professor in Sociology at the University of Exeter. Her work focuses on the use of forensic technologies in policing in England and Wales and the ways in which different stakeholders, including law enforcement and the judiciary, understand the role of forensic science in the criminal justice system. Her latest projects have been funded by the British Academy and the Economic and Social Research Council and have followed the development and application of digital forensics in crime investigations. Dana's work has been published in criminology, sociology and policing journals, including the *British Journal of Criminology*, *Policing and Society* and *Sociology*.

Introduction

1

FORENSIC GENETICS AND THEIR TECHNOLEGAL WORLDS

Victor Toom, Matthias Wienroth and Amade M'charek

In April 2018, Joseph DeAngelo, the Golden State Killer, was spectacularly identified as a long-sought serial offender of serious crimes dating from 1974 to 1986. The former police officer was accused of at least 13 murders, 50 rapes and 120 burglaries committed during that time. The case had gone 'cold' because, despite years of intensive investigations, including billboards showing composite drawings of a suspect, insufficient intelligence prevented detection of the perpetrator. The state of California even introduced new legislation in 2002, allowing it to set up a forensic DNA database in the hope of solving these crimes. Although this database did help establish a link between these crimes, it did not provide clues about the identity of the perpetrator.

DeAngelo eventually became a suspect via the use of a publicly accessible DNA database. Police investigators worked with a genealogist who helped them compare the Golden State Killer's DNA profile, generated from crime scene traces, with profiles in recreational and commercial DNA databases. The idea was to identify the suspect through genetic genealogical research. Comparing crime scene DNA to profiles already uploaded to databases such as GEDmatch, the police investigators identified distant relatives of the hitherto-unknown suspect and compiled the family trees of some 25 men who shared the same great-great-great-great-grandparent. Two such trees led to two men who were identified as suspects while the rest could be excluded based on appearance, ethnicity or place of residence. One suspect was traced and could be ruled out with the help of DNA comparison. In the case of the second suspect, the police secretly collected traces. This DNA yielded a complete match.

This use of recreational and commercial DNA databases has sparked a wave of excitement within the forensic world (e.g., Greytak et al. 2019; Katsanis 2020; Kennett 2019; Kling et al. 2021). More than 200 cold cases have already been resolved in the USA using forensic genetic genealogy. When this method was

DOI: 10.4324/9780429322358-2

applied in a long-running Swedish double homicide investigation in June 2020 and the suspect was arrested through a comparison with two North American ancestry DNA databases, GEDmatch and FamilyTreeDNA (Tillmar et al. 2021), discussion about the application of forensic genetic genealogy gained momentum in other European countries as well. While the potential is evident, many concerns about this technology exist (e.g., Granja 2021; Murphy 2018; Samuel & Kennett 2020). For example, medical and forensic DNA databases are subject to clearly stipulated (legal) rules concerning collection, processing, storage and, in the case of forensic databases, also for the destruction of DNA material and profiles (e.g., Skinner & Wienroth 2019). Recreational and commercial databases, however, do not share such legal guidelines because they were not set up for medical or forensic purposes. Although their DNA data is subject to informed consent, many questions remain open, including about the ownership of genetic materials and data (e.g., when databases are transferred or sold), the re-purposing of data (e.g., for criminal investigations) and changes in user policies (e.g., when GEDmatch and FamilyTreeDNA opened their customers' profiles to criminal investigators).

The example of the Golden State Killer lands us in the thick of the plot of this volume. While the technology of forensic genetic genealogy is emergent, it provides an exemplary case of the ways forensic DNA operates in the world and how it relates to or challenges the law. Rather than a confined add-on to criminal investigation, forensic DNA is a potent actor that co-shapes worlds in and outside the field of criminal justice. Since its introduction in the mid-1980s, it has reshaped investigative policing in non-trivial ways (Williams and Johnson 2008; M'charek 2008; Toom 2011). The case of the Golden State Killer underscores the ways in which genetic identification, or the potential thereof, is impacting societies way beyond the criminal justice system. Although recreational and commercial DNA databases have been drawn on for crime solving in the USA for some years now, other countries are still probing the legal and ethical challenges of this technology (e.g., M'charek & de Knijff 2021; in this volume, see Lawless; Murphy). This volume charts some of the complex relationships between forensic genetics technologies, the law and society, reflecting on what in this book we call the 'technolegal worlds' of this technoscience.

Forensic DNA profiling, arguably, was first deployed by Sir Alec Jeffreys and colleagues in the mid-1980s in a family reunion case and later in the investigation of the Colin Pitchfork case in the United Kingdom. Its potential for criminal investigations was relatively swiftly acknowledged and led to the implementation of this technology in many national jurisdictions throughout the 1990s and 2000s (Hindmarsh & Prainsack 2010). With increasing standardisation efforts since the mid-1990s, DNA profiling and databasing have come to be regarded as some of the most reliable tools – framed in terms of 'gold standard' and 'truth machine' (Lynch 2003, Lynch et al. 2008) – for achieving criminal justice purposes such as detection, prosecution and prevention. These DNA technologies have become so widely known and reproduced in public discourse that, despite the relatively rare use of DNA in criminal justice, genetic essentialism (DNA must be involved) and genetic exceptionalism (DNA has unique power to identify and inform) have engendered the emergence of a forensic imaginary for forensic genetics (Williams 2010).

Yet forensic DNA was not just an addition to pre-existing practices, nor did it simply travel from the laboratory to the criminal justice system (e.g., Hindmarsh & Prainsack 2010; M'charek 2008; M'charek 2017; Toom 2011). Rather it has produced *technolegal worlds*: assemblages of technologies and legal objects, with their associated standards and norms, that co-create novel identities, practices and sociotechnical innovations. Forensic genetics technologies have been examined by scholars in science and technology studies (STS), socio-legal studies (SLS) and law, attending to the interactions and entanglements between science and the law in forensic settings (e.g., Amelung et al. 2021; Chow-White & Duster 2011; Duster 2006; Jasanoff 1998; Lynch et al. 2008; Machado & Granja 2022; McCartney 2006; Williams & Johnson 2004, 2008; and many more we cannot name here, including the contributors to this volume). The aim of this volume is to extend the scope of this scholarship by attending to forensic DNA profiling in countries that have been neglected thus far and by analysing novel and emerging technological innovations that anticipate, and are dependent on, a translation of social practices. We offer an analysis of forensic genetics through the lens of the concept 'technolegal worlds'.

Technolegal worlds

Drawing on Donna Haraway's (1997) take on the notion of 'technoscience', we consider forensic science and technology as social and material practices: they are inscribed with cultural values through the entanglement of physical aspects with symbolic practices, which, as the saying goes, *are made but not made up*. Technoscience, therefore, is not simply a tool. Rather, it is practice in which materiality and (semiotic) meaning co-configure each other; where subjects and objects are active, affective and affected; and where lifeworlds and social horizons are constantly remade. Forensic practices, made of things and people, laws and DNA entangled in heterogeneous networks, constitute and perform *articulate collectives* (M'charek 2008; 2017). Here, scientific knowledge and technologies enable and facilitate, among other things, investigations in criminal justice systems; forensic operations after disaster (Toom 2016, 2018a; Williams and Wienroth 2014a, 2014b), humanitarian crisis (M'charek & Casartelli 2019) and human rights violations (Schwarz-Marin and Cruz-Santiago 2016; Toom 2020; García-Deister & Smith 2020); and health, commercial, and genealogical research via databases (Nelson 2008; Panofsky & Donovan 2019). The resulting practices and subsequent utility of forensic genetics, differing across jurisdictions, come with specific sets of politics, norms, values and resolutions that contribute or challenge the legitimacy of using forensic technologies. Some of these practices (re)create and (re)consolidate individuals and populations: e.g., in creating new suspect identities (McCartney 2004; Lynch & McNally 2009; Toom 2012) or in how people are related through a process of 'racialisation' (M'charek 2020; Bartram et al. 2022; M'charek & Wade 2020). In practices where forensic genetics is deployed to find and identify persons, new understandings about the type of relationship that is (re)produced – between those looking and those 'being found' – may be foregrounded. For example, if a person submits their DNA to a company providing genealogical services, they may create the perception of 'family' by producing a genetic unity of individuals who otherwise share no social familial bonds simply because they have a common ancestor six generations back (Haimes & Toom 2014). This genetic genealogy often constitutes recreational use of genetics but may also be used for hereditary health reasons or in forensic casework, as discussed in the opening of this chapter. In the case of disaster victim identification (DVI), where victims' remains are to be identified through forensic science, including genetics, authorities must act scientifically and legally yet also have to accommodate surviving family members' emotions like grief, mourning or relief (Toom 2018b). In, both, recreational use of genetics and DVI, DNA profiling and databasing relate to, and interfere with, the architecture of security, including the fact that identifying persons may contribute to legal issues like solving cold cases and connecting human remains to a missing person. Thus, forensic technoscience foregrounds the complexities of notions such as the individual, the collective, ancestry and security, as well as their interactions in material practice.

The deployment of forensic genetics for security and justice purposes yields wider social consequences. For example, serious legal costs might arise when matching a suspect's DNA profile with one obtained from a trace lifted at a crime scene. Such a match necessarily implicates the subject in an ongoing criminal investigation, but the investigative work is not finished. A match alone is usually insufficient to prosecute. Investigative authorities are tasked with finding further evidence to compellingly demonstrate that a 'genetic suspect' is responsible for a particular crime, linking source and action level analyses (e.g., Toom 2011; M'charek et al. 2013). While a DNA match may provide a strong clue of someone's proximity to a crime scene, perhaps even their involvement in a crime, such a mechanism also comes with risks: e.g., authorities may be convinced of a person's involvement before sufficient evidence is provided (see reasonable doubt). Tunnel vision is a well-known mechanism producing miscarriages of justice, and a DNA match may easily contribute to investigators' and adjudicators' views that a person must be the perpetrator. In such a circumstance, a suspect may be motivated to engage with law enforcement authorities, providing information they may otherwise consider too personal to share with strangers. If this leads to the suspect having to explain how their DNA may have ended up at a crime scene, it stands in contrast to the 'onus of proof' on the prosecution's side, perhaps even infringing on the liberal-democratic value of 'innocent until proven guilty' (M'charek et al. 2012; Toom et al. 2016). While these are not reasons to argue against DNA profiling and databasing, these considerations hint at the potentially complicated relationship between subjects living in a state of law and the incriminating powers of forensic genetics. Thus, from a liberal-democratic perspective, the deployment of forensic genetics technologies requires active and reflective work (Jasanoff 2003; Lynch & McNally 2009), including, e.g., transparent validation of technologies and their use and standardised and tested practices, including accreditation, proportionate legislation and regulation, to name but a few key aspects for legitimacy (Wienroth 2020a) in liberal-democratic technolegal worlds. These are vital to safeguard the rights of individual citizens, communities and the wider resident population.

Simultaneously, forensic genetic practices are not only about legal issues; they are also deeply political. Decisions have to be made about which technologies to deploy, how to validate them, what reference data to draw from and who – police, commercial, academic labs, accredited or not – can provide data and services in the criminal justice system and for other security and justice purposes. Policy decisions are also made on, e.g., the categories of persons to be included in forensic DNA databases: from convicted offenders of serious crime, via the inclusion of volume crime offenders (or even political prisoners in authoritarian regimes) right down to expansive criteria that include migrants (e.g., via databases that are also used forensically), arrestees, suspects, persons of interest and volunteers (see Hindmarsh & Prainsack 2010; Richter and Louzada, this volume). Political and legal decisions are made about the storage of DNA samples and data, including on who oversees database uses and on retention practices (Skinner & Wienroth 2019). Forensic genetic practices also change how police authorities organise investigations, e.g., suspects or persons of interest are routinely produced simply by partially or fully matching DNA profiles of individuals on databases to profiles lifted from traces. Furthermore, forensic practices in policing and elsewhere increasingly draw together various entities previously not or less apparently connected. Examples are the exchange of DNA data through the Prüm Decisions (see later in this chapter for more information) or because specific traits may become markers for suspicion. As a result, forensic genetics technosciences are part of wider political economies in which diverse interests and actors come together.

This volume collects a series of analyses of *technolegal* issues, referring to forensic genetics as a technoscientific practice that is social and material and a legal practice that is also deeply political. The neologism *technolegal* invites authors and readers to attend to the materialities of forensic discourses and practices; it moves beyond a sole focus on forensic technology and investigation as technical practice to consider the wider norms, values, interests, legitimacy and resolutions that co-constitute the political economies of forensic technoscience. The term *forensic* originates from the Latin *forensis* and is etymologically related to *forum*, denominating a public market or square. Especially in the Roman era, the forum was not only a place where goods were traded, but also, most importantly, a place where citizens would gather for public matters, including (political) decisions, arbitration and adjudication, as well as the execution of sentences. These practices of mobilising the public, but also of the wider political economy of forensic genetics, are discussed in this volume. We acknowledge that the attributive concept *technolegal* is broad and argue that we can use it best as a methodological tool rather than strict nomenclature. The concept facilitates the opening up of analytical possibilities (cf. Mol 2013).

While we aim to keep the concept 'technolegal' methodologically and conceptually open, 'under development' as it were, we are aware of its diverse disciplinary and conceptual origins (see Lynch et al. 2008; Toom 2016, 2020; Wienroth 2018).

One important genealogy is a social scientific discipline often termed *science and technology studies* or *science, technology and society* (both abbreviated STS). This interdisciplinary field of STS has two important branches to look at science in society that are relevant for us here: actor-network theory (ANT) and sociology of scientific knowledge (SSK). ANT is best known for its radical approach to the ontological status of phenomena. Things, human or non-human, are understood as networks of relations, rather than singular entities that are contained in themselves. As these networks of relations are loosely connected and constantly shifting, the identity of things is not stable. An ANT approach thus attends to the fluidity of things and to the active work that goes into stabilising them, offering a methodological sensitivity for studying and relating to the world around us. Thus, the denominator 'technolegal' includes many different modalities and realms that come to matter in the analysis of forensic genetic practices (see Toom 2016). Equally, other STS scholars, e.g., colleagues drawing from SSK (see Kruse 2015; Lawless 2016; Lynch et al. 2008), may point to the analytical necessity for a demarcation between the technological and the legal in order to understand the way that knowledges are (re) created, (re)consolidated and, in their co-production, create *socio-technical* orders. SSK and related scholarship in, e.g., the social studies of forensic science, attend to the formation of communities – judges, police detectives, forensic scientists or technicians – and how they develop discourses and practices and make decisions (e.g., Jasanoff 1995). Value here lies in analysing a diversity of perspectives and accounts as community-specific interpretations of an issue, thus situating issues within epistemological frameworks to analyse how such issues are resolved time and again (Lynch et al. 2008). The status of a claim – for example, the value of a match and whether that incriminates a suspect – is decided through a process of closure of the controversy in which power, credibility and assumptions all play their part (Jasanoff 1998; Lynch 1998). Controversies are resolved through interactions between social groups; therefore, phenomena that are analysed do not dictate the outcome (Collins 1981). In such an approach, a dualism enables analysis of technoscience and the law. In this volume on technolegal worlds, analytical co-habitation is accomplished through keeping apart stylistically the realms of technoscience as well as the law and the legal – *technolegal* (e.g., Wienroth 2018).

This volume sets the *technolegal* in conversation with *worlds*. The notion of *worlds* refers to various issues the chapters in this book aim to capture. The first and most apparent is that of geography and scale, that of global science and local technolegal rules: authors from different parts of the world have been invited to analyse local and global engagements with forensic genetics technoscience. Such analyses are urgently required. Williams and Johnson (2008) demonstrate that many jurisdictions, when introducing forensic DNA profiling and databasing, follow a common trajectory from rarely deploying technology (e.g., only in investigating major violent crimes) to routinely using them (e.g., in volume crime). Yet some countries that only recently introduced these technologies may follow aspects of this trajectory and depart on others (cf. Wienroth et al. 2014). Examples include Portugal's database, which only includes convicted offenders (not suspects or traces) and re-draws boundaries in its participation of Prüm (Amelung & Machado 2019); efforts of the Chinese government to include the complete Uighur Muslim community in criminal DNA databases (Cyranoski 2017; Moreau 2019); and discussions about including complete national populations in a DNA database (Dedrickson 2017; Kaye & Smith 2003; Toom 2014). We are very glad to have gained contributions from scholars doing research on technolegal practices in countries which have been understudied in the field of forensic science studies: Brazil, Colombia, Ghana and South Africa (see contributions in Part I of this volume).

In the social sciences, the notion of 'world' has further connotations. Worlds are inhabited by subjects and objects who (re)produce them. Consequently, worlds are emergent, and the processes of (re)production can be conceptualised as the descriptor *worlding*. The concept has been coined and furthered in several disciplines, including feminist philosophy, postcolonial studies, international relations, urban studies and STS. Where Spivak (1985), for example, refers to worlding as a process whereby the worlds of native people are transformed into 'colonized spaces' under colonial rule, urban studies scholars deploy worlding to 'identify the projects and practices that instantiate some vision of the world in formation' (Ong 2011: 11). In both instances, the lived world is not a stable given, but subject to ontological change. This understanding is also exemplified in Kenney's (2015: 764) theoretical discussions about relational empiricism, where she argues that worlding is a descriptor of activities that are locally situated and, as such, normative and political. In this volume, diverse processes of worlding, through and in technolegal practices in different parts of the globe, are at the centre of the analysis. The focus on worlds and worlding is yet another invitation to chapter authors to be attentive to the specific, contextual politics and normativities of forensic sciences – the good, the bad and the nuanced. This applies to the various jurisdictions discussed in the volume (Part I), to emerging technologies in criminal investigation (Part II) and to the many issues in the context of legitimacy, including social acceptability (Part III). Before we provide an outline of the book, we want to provide a short introduction to forensic genetics developments in order to set the scene for the book's chapters.

A brief genealogy of forensic genetics in four waves and technolegal worlds

Forensic genetics has emerged in overlapping waves of sociotechnical innovation (Wienroth et al. 2014). Each of these waves can occur at different times and places as they are not meant to represent a linear history but rather an understanding of the emergence of sociotechnical innovations. These innovations have created in their wake diverse technolegal worlds made up of technologies and their uses, oversight and negotiation.

The first wave is associated with Jeffreys and colleagues' discovery of 'DNA fingerprinting' (Gill et al. 1985). The technology developed at the time was dependent on large traces of biological material with high concentration of DNA (e.g., blood and semen). The typing of a sample could take weeks, and comparing various profiles was difficult as standards across different laboratories had yet to be developed (Lander 1989). Because of this lack of harmonisation in practices, significant controversies arose, including around the proper handling of crime scene traces (Lander 1989; Lynch & Jasanoff 1998; Lynch et al. 2008) and the correct statistical interpretation of a match (Lewontin & Hartl 1991; Lynch et al. 2008; M'charek 2000). These controversies were discussed in court and in scientific journals, creating a scientific knowledge machine (Cole 2001; Jasanoff 1995; Lynch 1998). The technolegal world of this first phase as observed in the USA and in the UK during the 1980s and 1990s – the so-called DNA wars (Reibstein & Foote 1995) – was one of contention and doubt and the slow realisation that standards and oversight are vital to reliable and helpful forensic uses of DNA. Socio-ethical concerns focused on evidentiary power and utility for police work and in court.

Some of these issues were resolved during the second wave, when other DNA typing technologies (e.g., short tandem repeats, STRs) and methods to multiply biological material from samples with low quantity of DNA (polymerase chain reaction [PCR]) became standard in many forensic genetic labs (M'charek 2008; M'charek et al. 2012). Relatedly, and significantly, this second wave led to the introduction of DNA databasing, accompanied by dedicated forensic DNA legislation and regulation in countries in the 'global north' (Hindmarsh & Prainsack 2010; Krimsky & Simoncelli 2011; McCartney 2006; Skinner 2013; Williams & Johnson 2008). STR profiles are highly standardised and enable the easy storage and comparison of profiles stored in DNA databases (McCartney et al. 2011; Prainsack & Toom 2013). Accompanying this major step-change in forensic DNA have been increased professional, policy and public debates. These have emerged about new identities created through DNA profiling and databasing: who is a legitimate subject in a database, for how long should profiles and samples be retained, who should be able to access which databases and to what purposes should profiles and samples be put (Williams & Johnson 2004; Skinner 2020a; Skinner & Wienroth 2019). Furthermore, databasing has enabled the exchange of information across borders and jurisdictions; it has become a staple of forensic genetics in the European Union and in various bi- and tri-lateral agreements in the global North. More recently, we have been able to observe the use of different types of databases, as discussed in the opening of this chapter. Recreational, commercial and forensic genetics research databases have proven to retain valuable information for investigative authorities. Many customers now submit DNA samples to genetic analysis companies for profiling as well as health and ancestry analysis. Some upload their personal DNA profiles to dedicated genealogy databases like GEDmatch and FamilyTreeDNA in order to get a sense of their place in the world. GEDmatch (now owned by forensic service provider Verogen) and its competitors provide detailed information about, for example, ancestry and family trees (De Groot et al. 2021; De Groot, van Beers et al. 2021; Katsanis 2020). In recent years, samples obtained from unsolved crimes (often referred to as cold cases) have been submitted to those databases, resulting in the apprehension of persons thought to be responsible for committing those crimes (see Murphy, this volume). The technolegal worlds of the second wave bring together sociotechnical innovations of new technologies, formalised instruments of oversight, good practice and debates around, for example, privacy and equality, creating far more complicated and mutable technolegal assemblages than those of the first wave.

There is, however, one issue of this second wave which we would like to spend some more time exploring because it reflects on an emerging technolegal world of internationalisation and sets the scene for why we consider it not only important, but vital to look beyond the global North for an analysis of technolegal worlds of DNA profiling and databasing. Forensic DNA databases typically retain profiles of samples obtained from crime scenes and known individuals within a particular territory (e.g., state, country). But whereas criminals are not bound to a state or country's borders per se, the use of national DNA databases had been restricted to a particular jurisdiction. Over the last two decades, various initiatives have been implemented to address technical and legal limitations to cross-border uses. The most significant development are the so-called Prüm Decisions in the European Union (EU) and some associated states (Machado & Granja 2020; Machado et al. 2020). These decisions date back to the mid-2000s, when some EU member states decided to commence sharing DNA data held on their national databases. The EU Council considered this project valuable and transposed the 2005 Prüm Convention into the body of binding common EU rights and obligations in 2008 (Prainsack & Toom 2013). This decision had two significant effects. First, it rendered the establishment of national DNA databases in the EU area mandatory. Consequently, several EU member states (e.g., Italy, Ireland, Greece) were obliged to develop and implement forensic DNA legislation aimed at the governance of a national DNA database. Second, the Prüm Decisions contributed to a network of national DNA databases that exchange DNA data daily, aimed at identifying unknown crime suspects or combating transnational crimes. While the exchange of DNA data in the context of the Prüm Decisions helps to trace suspects, it is unknown how effective the system is as a crime-fighting tool. It is therefore impossible to assess whether the Prüm Decisions and their regime are proportional crime-control measures (Toom 2018a, 2018b; Toom et al. 2019; see also McCartney and Amankwaa, this volume). It is in that context that a further expansion of the Prüm Decisions, the so-called next generation Prüm, by allowing more (non-EU) countries into the network of national databases and adding new technologies (e.g., facial recognition) has received critiques from scholars and organisations involved in the protection of basic rights (Machado et al. 2022; Toom et al. 2019).

The third wave of forensic genetic innovations has crested with further technological developments around testing for and inferring the appearance of an unknown person, as well as their biogeographic ancestry. Existing and new forensic markers have been identified and deployed and are in the process of being associated with certain appearance traits. A consequence of these developments is that forensic genetics no longer brings only individuals into the view of investigative authorities, but also categories of traits and populations (Cole & Lynch 2006; Jong & M'charek 2018; M'charek 2008; Toom 2012; Toom et al. 2016). Such 'suspect populations', based on a shared characteristic like skin, hair and eye colour or geographic ancestry, have the potential to render entire communities suspect via technoscience. Indeed, forensic genetics practices in this field strongly resonate with notions of race and ethnicity (M'charek 2020; Hopman & M'charek 2020; Khan & Machado 2022; Skinner 2020b; Van Oorschot & M'charek 2021). While the technolegal worlds of the first wave of forensic genetics primarily focused on scientific and operational (police and court) debates, the second and third waves have expanded these to policy debates. In the third wave, public, policy and practitioner engagement are taking place with legitimacy around the introduction and uses of these new technologies at their heart (e.g., Lipphardt et al. 2016, 2019; Wienroth 2018, 2020a).

The fourth wave refers to the introduction of 'massive parallel sequencing' (MPS) as an expansion of different types of forensic analyses as well as their simultaneous analysis. MPS brings together the promise of the most significant elements of the first three waves and goes beyond those by challenging the boundaries of the forensic and medical but also other domains (such as recreational ancestry testing or the commercial world of forensic services). Samples are not just expected to be sequenced more quickly, more efficiently and with increased sensitivity and more information compared to individual matching (first wave) and attribution (third wave). MPS is a change in the philosophy of DNA profiling and databasing, potentially leading to developments around lifestyle testing via forensic (epi) genomics (e.g., Vidaki & Kayser 2017). In the meantime, MPS saw its premiere in a Dutch legal case in 2019 (Hopman et al., this volume) and has been deployed in forensic casework by the International Commission on Missing Persons (Parsons et al. 2019). The technolegal worlds of the fourth wave are emerging and highly mutable; significant work is being done by scholars to understand how forensic practice from lab to societal impact can be shaped to be societally more robust and ethically sound (e.g., Wienroth 2020b; Wienroth et al. 2021).

Chapter overview

This volume is divided in three parts. Each part discusses relevant aspects of technolegal worlds and addresses issues commonly understudied in the scholarly literature on forensic genetics. Here we discuss the common themes of each part and introduce the chapters.

In Part I, case studies about the governance, legal mechanisms and exemplary cases for a variety of countries are introduced. This section includes a selection of jurisdictions across the globe, including Brazil, Colombia, Ghana and South Africa. There is only limited knowledge currently available in print about these jurisdictions and how they govern DNA profiling and databasing. Taking Colombia as their geographical unit of analysis in Chapter 2, authors Olarte-Sierra and Castro Bermúdez analyse the social, scientific and political disputes that emerge when attempts are made to identify the victims of a violent conflict stretching out over decades. Positioning the identification work in narratives about conflict-related violence, their analysis shows that a focus on technolegal worlds enriches the understanding and nuances of, in their case, the armed conflict in Colombia. The subsequent chapter in the volume, written by Richter and Louzada, addresses the forensic DNA database in Brazil. They show that the establishment of the Brazilian national database is partly the result of a biotech lobby and the strong involvement of forensic scientists. Nevertheless, despite these two strong incentives to create a national DNA database, the Constitution of Brazil together with the country's past violent dictatorship stand in the way of subject bodies being sampled for DNA analysis. Chapter 4, by Amankwaa and Amankwa Addo, narrates the challenges that the West African country Ghana deals with in attempting to implement a technolegal world of forensic genetics. The chapter demonstrates the work that goes into establishing a forensic DNA database, including legal issues, standardisation, and decision-making about governance and governance principles. Their analysis ends with a reflection on the shortcomings still to be addressed, arguing that the 'technolegal world in Ghana is unregulated and tenuous'. Chapter 5 by Tamarkin focuses on the emergence of the South African forensic DNA database. The history of this database cannot be understood without reference to South Africa's apartheid system. Tamarkin narrates this history of social and political contestation through a focus on parliamentary debates and describes how legacies of racial oppression helped shape the emergence of the database.

In Part II, innovative technologies and the technolegal worlds they help establish are analysed: massive parallel sequencing, rapid DNA, forensic DNA phenotyping and ancestry analysis. While most of these applications have received attention in the body of literature, their technolegal politics have not been discussed so far in one volume that simultaneously discusses other novel approaches and their practices. Each of the three chapters in Part II analyses state-of-the-art technologies and their practices and simultaneously provides ample opportunity to further the notion of technolegal politics. One of the latest technolegal developments – massive parallel sequencing (MPS) – is analysed by Hopman, van Oorschot and M'charek in Chapter 6. Their focus regards partly invisible work by legal and forensic actors aimed at making ready the Dutch judiciary for MPS. More specifically, they follow a case of rape that occurred in the Netherlands in which MPS was, for the first time, accepted as legal evidence in a court of law. In Chapter 7, Wilson-Kovacs also focuses, albeit discursively, on the work required to have adopted another technology. This technology can type a given sample in less than two hours and therefore is considered superior in efficiency, value and effectiveness. Regarding what is usually referred to as rapid DNA, Wilson-Kovacs situates her analysis in the English criminal justice system. She shows how the promise of rapid DNA brought Home Office, police personnel and commercial providers together in attempts to make the case for this technology. A systematic analysis of rapid DNA analysis, forensic DNA phenotyping and forensic genealogy is provided by Lawless in Chapter 8. For each of these technologies, he provides a 'social realist' account by reflecting on the social benefits, epistemic caution and social risks. While using his analysis to elaborate on new technolegal worlds, he also reflectively asks whether one of the most successful concepts in STS, 'co-production', is by and in itself a sociotechnical imaginary.

Part III discusses various issues of legitimacy in relation to forensic genetics and their technolegal worlds. We define *legitimacy* rather broadly in this volume, including legal and social connotations. The chapters elaborate the acceptability and desirability of developments by addressing trust, temporalities, effectiveness, solidarity and conceptions such as consent and family in and of technolegal worlds. Each chapter is committed to articulating technolegal politics of the newly emerging and (de)stabilising worlds it describes. Fundamental to forensic genetics as a technology in the criminal justice system is trust. Trust is not given but actively construed and negotiated. In addition, trust is not just one thing but comes in many guises. In Chapter 9, Wienroth analyses trust in forensic genetics and identifies three core relationships based on epistemic trust, operational confidence and courtroom credibility. In his analysis, Wienroth suggests that 'systemic trust' built on 'technologies of situated trustworthiness' constitutes trust in forensic genetics technolegal worlds. Taking the example of DNA testing in family reunification, in Chapter 10, Helén and Tapaninen draw on their empirical work in Finland. They attend to the entanglement of legislation, administration and biotechnology in family reunification cases. Instead of addressing the problems of genetic relatedness and biological kinship, the analysis presented by the authors puts temporality at the forefront. More specifically, they address the time of legislation and time of technoscience as drivers for technolegal regimes. Chapter 11 engages with forensic DNA databases. While they deliver invaluable information as they connect, through time and place, crime scene samples to each other and to subjects included in a database, it remains unclear how efficient and effective databases actually are in contributing to solving crimes. McCartney and Amankwaa propose a model that puts the notion of the 'integrity' of DNA databases centre stage. Interestingly, this concept denotes something both 'whole' and 'ethical' and includes the key notions of viability, legitimacy and acceptability. Chapter 12 introduces the concept of solidarity to discussions of technolegal worlds in forensic genetics using the example of forensic DNA phenotyping (FDP). Prainsack and Samuel argue that, while FDP risks discriminating against ethnic minorities and marginalised groups, the predictive analysis of an unknown suspect's possible visible characteristics and biogeographical ancestry can help investigations. They suggest that the principle of solidarity may help overcome the dichotomy between the individual and the collective in technolegal worlds and offer suggestions to implement a solidarity-infused understanding of security for FDP.

This volume started with a case being solved through forensic genetic genealogical research. In Chapter 13, Murphy analyses three taken-for-granted categories in this type of forensic research: consent, family and jurisdiction. Her analysis clearly demonstrates that there is a misfit between such notions and their language and the emerging technolegal worlds enacted through genealogical research.

The volume closes with an epilogue by Skinner on truth-making and genetic citizenship. While reflecting on the chapters, he reiterates that truth, while usually considered universal, is produced in multiple ways in which technologies, legal systems and local circumstances all play their part. Referring to the notion of 'bio-citizenship', Skinner asks us to consider how forensic practices help enroll particular persons into new configurations that further injustices and disparities.

As the contributions to this book show, technolegal worlds are far from innocent; they produce new subjectivities and objectivities, and they remain mutable. Rather than defining the concept statically, our aim in introducing the concept of technolegal worlds is to provide a critical tool to scholars in the fields of forensic genetics and other technosciences of justice and security with a framework sufficiently flexible to combine empirical insight with theoretical rigour.

References

Amelung, N., Granja, R. and Machado, H. (2021) *Modes of Bio-bordering: The Hidden (dis) Integration of Europe*. Singapore: Springer Nature.

Amelung, N. and Machado, H. (2019) '"Bio-bordering" processes in the EU: De-bordering and re-bordering along transnational systems of biometric database technologies', *International Journal of Migration and Border Studies*, 5(4), pp. 392–408.

Bartram, I., Plümecke, T. and Schultz, S. (2022) 'Genetic racial profiling: Extended DNA analyses and entangled processes of discrimination', *Science & Technology Studies*, 35(3), pp. 44–69. doi: 10.23987/sts.101384.

Chow-White, P. A. and Duster, T. (2011) 'Do health and forensic DNA databases increase racial disparities?' *PLoS Medicine*, 8(10), p. e1001100.

Cole, S. A. (2001) *Suspect Identities. A History of Fingerprinting and Criminal Identification*. Cambridge, MA and London: Harvard University Press.

Cole, S. A. and Lynch, M. (2006) 'The social and legal construction of suspects', *Annual Review of Law and Social Science*, 2, pp. 39–60.

Collins, H. M. (1981) 'Son of seven sexes: The social destruction of a physical phenomenon', *Social Studies of Science*, 11(1), pp. 33–62.

Cyranoski, D. (2017) 'China expands DNA data grab in troubled western region', *Nature*, 545(7655).

De Groot, N. F. *et al.* (2021) 'Accessing medical biobanks to solve crimes: Ethical considerations', *Journal of Medical Ethics*, 47(7), pp. 502–509.

De Groot, N. F., van Beers, B. C. and Meynen, G. (2021) 'Commercial DNA tests and police investigations: A broad bioethical perspective', *Journal of Medical Ethics*, 47(12), pp. 788–795.

Dedrickson, K. (2017) 'Universal DNA databases: A way to improve privacy?' *Journal of Law and the Biosciences*, 4(3), p. 637.

Duster, T. (2006) 'Explaining differential trust of DNA forensic technology: Grounded assessment or inexplicable paranoia?' *The Journal of Law, Medicine & Ethics*, 34(2), pp. 293–300. Available at: http://dx.doi.org/10.1111/j.1748-720X.2006.00034.x.

García-Deister, V. and Smith, L. A. (2020) 'Migrant flows and necro-sovereignty: The itineraries of bodies, samples, and data across the US-Mexico borderlands', *BioSocieties*, 15(3), pp. 420–437.

Gill, P., Jeffreys, A. J. and Werrett, D. J. (1985) 'Forensic application of DNA "fingerprints"', *Nature*, 318(12), pp. 577–579.

Granja, R. (2021) 'Long-range familial searches in recreational DNA databases: Expansion of affected populations, the participatory turn, and the co-production of biovalue', *New Genetics and Society*, 40(3), pp. 331–352.
Greytak, E. M., Moore, C. and Armentrout, S. L. (2019) 'Genetic genealogy for cold case and active investigations', *Forensic Science International*, 299, pp. 103–113.
Haimes, E. and Toom, V. (2014) 'Hidden in full sight: Kinship, science and the law in the aftermath of the Srebrenica genocide', *New Genetics and Society*, 33(3), pp. 277–294. Available at: www.tandfonline.com/doi/abs/10.1080/14636778.2014.946004.
Haraway, D. J. (1997) *Modest_Witness@Second_Millennium.FemaleMan(c)_Meets_OncoMouse (TM). Feminism and Technoscience*. New York: Routledge.
Hindmarsh, R. and Prainsack, B. (2010) *Genetic Suspects. Global Governance of Forensic DNA Profiling and Databasing*. Edited by R. Hindmarsh and B. Prainsack. Cambridge: Cambridge University Press.
Hopman, R. and M'charek, A. (2020) 'Facing the unknown suspect: Forensic DNA phenotyping and the oscillation between the individual and the collective', *BioSocieties*, 15(3), pp. 438–462.
Jasanoff, S. (1995) *Science at the Bar. Law, Science, and Technology in America*. Cambridge, MA: Harvard University Press.
Jasanoff, S. (1998) 'The eye of everyman: Witnessing DNA in the Simpson trial', *Social Studies of Science*, 28(5), p. 713.
Jasanoff, S. (2003) 'Technologies of humility: Citizen participation in governing science', *Minverva*, 41, pp. 223–244.
Jong, L. and M'charek, A. (2018) 'The high-profile case as "fire object": Following the Marianne Vaatstra murder case through the media', *Crime, Media, Culture*, 14(3), 347–363. https://doi.org/10.1177/1741659017718036.
Katsanis, S. H. (2020) 'Pedigrees and perpetrators: Uses of DNA and genealogy in forensic investigations', *Annual Review of Genomics and Human Genetics*, 21(1), pp. 535–564. doi: 10.1146/annurev-genom-111819-084213.
Kaye, D. H. and Smith, M. E. (2003) 'DNA identification databases: Legality, legitimacy, and the case for population-wide coverage', *Wisconsin Law Review*, p. 413.
Kennett, D. (2019) 'Using genetic genealogy databases in missing persons cases and to develop suspect leads in violent crimes', *Forensic Science International*, 301, pp. 107–117.
Kenney, M. (2015) 'Counting, accounting, and accountability: Helen Verran's relational empiricism', *Social Studies of Science*, 45(5), pp. 749–771.
Khan, S. and Machado, H. (2022) 'Postcolonial racial surveillance through forensic genetics', in *Racism and Racial Surveillance: Modernity Matters*. London: Routledge.
Kling, D. *et al.* (2021) 'Investigative genetic genealogy: Current methods, knowledge and practice', *Forensic Science International: Genetics*, 52, p. 102474.
Krimsky, S. and Simoncelli, T. (2011) *Genetic Justice. DNA Databanks, Criminal Investigations, and Civil Liberties*. New York: Columbia University Press.
Kruse, C. (2015) 'The social life of forensic evidence', in *The Social Life of Forensic Evidence*. Oakland: University of California Press.
Lander, E. S. (1989) 'DNA fingerprinting on trial', *Nature*, 339, pp. 501–505.
Lawless, C. (2016) *Forensic Science: A Sociological Introduction*. London: Routledge.
Lewontin, R. C. and Hartl, D. (1991) 'Population genetics in forensic DNA typing', *Science*, 254, pp. 1745–1750.
Lipphardt, V., et al. (2016) *Open Letter on Critical Approaches to Forensic DNA Phenotyping and Biogeographical Ancestry*. Iowa: Freiburg.
Lipphardt, V., Buchanan, N., Pfaffelhuber, P., Staubach, F. and Wienroth, M. (2019) 'Interdisziplinäre Überlegungen zu Erweiterten DNA-Analysen (Interdisciplinary

considerations about new and emerging forensic DNA analyses)', *Jahrbuch für Wissenschaft und Ethik*, 24(1), 119–154.
Lynch, M. (1998) 'The discursive production of uncertainty: The OJ Simpson "dream team" and the sociology of knowledge machine', *Social Studies of Science*, 28(5–6), pp. 829–868.
Lynch, M. (2003) 'God's signature: DNA profiling, the new gold standard in forensic science', *Endeavour*, 27(2), p. 93.
Lynch, M., Cole, S. A., McNally, R. and Jordan, K. (2008) *Truth Machine. The Contentious History of DNA Fingerprinting*. Chicago and London: The University of Chicago Press.
Lynch, M. and Jasanoff, S. (1998) 'Contested identities: Science, law and forensic practice', *Social Studies of Science*, 28(5–6), pp. 675–686.
Lynch, M. and McNally, R. (2009) 'Forensic DNA databases: The co-production of law and surveillance technologies', in Atkinson, P., Glasner, P. and Lock, M. (eds) *Handbook of Genetics and Society. Mapping the New Genomic Era*. London: Routledge.
M'charek, A. (2000) 'Technologies of population: Forensic DNA testing practices and the making of differences and similarities', *Configurations*, 8, pp. 121–158.
M'charek, A. (2008) 'Silent witness, articulate collectives: DNA evidence and the inference of visible traits', *Bioethics*, 22(9), pp. 519–528.
M'charek, A. (2017) 'Performative circulations: On flows and stops in forensic DNA practices', *TECNOSCIENZA: Italian Journal of Science & Technology Studies*, 7(2), pp. 9–34.
M'charek, A. (2020) 'Tentacular faces: Race and the return of the phenotype in forensic identification', *American Anthropologist*, 122(2), 369–380.
M'charek, A. and Casartelli, S. (2019) 'Identifying dead migrants: Forensic care work and relational citizenship', *Citizenship Studies*, 23(7), pp. 738–757.
M'charek, A. and de Knijff, P. (2021) 'Commerciële DNA-databanken: een mixed blessing of een bedreiging voor de forensische praktijk?' *Justitiële Verkenningen*, 47(3), pp. 9–19.
M'charek, A., Hagendijk, R. and Vries, W. D. (2013) 'Equal before the law: On the machinery of sameness in forensic DNA practice', *Science, Technology, & Human Values*, 38(4), pp. 542–565.
M'charek, A., Toom, V. and Prainsack, B. (2012) 'Bracketing off population does not advance ethical reflection on EVCs: A reply to Kayser and Schneider', *Forensic Science International: Genetics*, 6(1), pp. e16–e17. doi: 10.1016/j.fsigen.2010.12.012.
M'charek, A. and Wade, P. (2020) 'Doing the individual and the collective in forensic genetics: Governance, race and restitution', *BioSocieties*, 15(3), pp. 317–328.
Machado, H. and Granja, R. (2020) *Forensic Genetics in the Governance of Crime*. Singapore: Springer Nature.
Machado, H. and Granja, R. (2022) *Genetic Surveillance and Crime Control: Social, Cultural and Political Perspectives*. London: Routledge.
Machado, H., Granja, R. and Amelung, N. (2020) 'Constructing suspicion through forensic DNA databases in the EU. The views of the Prüm professionals', *The British Journal of Criminology*, 60(1), pp. 141–159.
Machado, H., Granja, R. and Amorim, A. (2022) 'Ethical challenges of merging criminal identification and civil identification within the Prüm system', *Forensic Science International: Genetics*, p. 102660.
McCartney, C. (2004) 'Forensic DNA sampling and the England and Wales national DNA database: A sceptical approach', *Critical Criminology*, 12, pp. 157–178.
McCartney, C. (2006) 'The DNA expansion programme and criminal investigation', *British Journal of Criminology*, 46(2), pp. 175–192.
McCartney, C., Wilson, T. J. and Williams, R. (2011) 'Transnational exchange of forensic DNA: Viability, legitimacy, and acceptability', *European Journal on Criminal Policy and Research*, 17(4), pp. 305–322.

Mol, A. (2013) 'Mind your plate! The ontonorms of Dutch dieting', *Social Studies of Science*, 43(3), pp. 379–396.
Moreau, Y. (2019) 'Comment: Crack down on genomic surveillance', *Nature*, 576, pp. 36–38.
Murphy, E. (2018) 'Law and policy oversight of familial searches in recreational genealogy databases', *Forensic Science International*, 292, pp. e5–e9.
Nelson, A. (2008) 'Bio science: Genetic genealogy testing and the pursuit of African ancestry', *Social Studies of Science*, 38(5), pp. 759–783.
Ong, A. (2011) 'Introduction: Worlding cities, or the art of being global', in Roy, A. and Ong, Aihwa (eds) *Worlding Cities: Asian Experiments and the Art of Being Global*. Oxford: Wiley-Blackwell.
Panofsky, A. and Donovan, J. (2019) 'Genetic ancestry testing among white nationalists: From identity repair to citizen science', *Social studies of science*, 49(5), pp. 653–681.
Parsons, T. J. *et al.* (2019) 'Large scale DNA identification: The ICMP experience', *Forensic Science International: Genetics*, 38, pp. 236–244.
Prainsack, B. and Toom, V. (2013) 'Performing the union: The Pr??m decision and the European dream', *Studies in History and Philosophy of Science Part C : Studies in History and Philosophy of Biological and Biomedical Sciences*, 44(1), pp. 71–79.
Reibstein, L. and Foote, D. (1995) 'Let the DNA wars begin.', *Newsweek*, 125(20), p. 55.
Samuel, G. and Kennett, D. (2020) 'The impact of investigative genetic genealogy: Perceptions of UK professional and public stakeholders', *Forensic Science International: Genetics*, 48, p. 102366.
Schwartz-Marin, E. and Cruz-Santiago, A. (2016) 'Forensic civism: Articulating science, DNA and kinship in contemporary Mexico and Colombia', *Human Remains and Violence: An Interdisciplinary Journal*, 2(1), pp. 58–74. Available at: http://dx.doi.org/10.7227/HRV.2.1.5.
Skinner, D. (2013) '"The NDNAD has no ability in itself to be discriminatory": Ethnicity and the governance of the UK National DNA Database', *Sociology*, 47(5), pp. 976–992. Available at: http://soc.sagepub.com/content/47/5/976.abstract.
Skinner, D. (2020a) 'Race, racism and identification in the era of technosecurity', *Science as Culture*, 29(1), pp. 77–99.
Skinner, D. (2020b) 'Forensic genetics and the prediction of race: What is the problem?', *BioSocieties*, 15(3), pp. 329–349.
Skinner, D. and Wienroth, M. (2019) 'Was this an ending? The destruction of samples and deletion of records from the UK police national DNA database', *BJHS Themes*, 4, pp. 99–121.
Spivak, G. C. (1985) 'The Rani of Sirmur: An essay in reading the archives', *History and Theory*, 24(3), pp. 247–272.
Tillmar, A. *et al.* (2021) 'Getting the conclusive lead with investigative genetic genealogy—A successful case study of a 16 year old double murder in Sweden', *Forensic Science International: Genetics*, 53, p. 102525.
Toom, V. (2011) *Dragers van Waarheid. Twintig Jaar Forensisch DNA-onderzoek in Nederland [Carriers of Truth. Twenty Years of DNA Profiling in the Netherlands]*. Deventer: Kluwer Press.
Toom, V. (2012) 'Bodies of science and law: Forensic DNA profiling, biological bodies, and biopower', *Journal of Law and Society*, 39(1), pp. 150–166.
Toom, V. (2014) 'Trumping communitarianism. Crime control and forensic DNA typing and databasing in Singapore', *East Asian Science, Technology and Society: An International Journal*, pp. 273–296.

Toom, V. (2016) 'Whose body is it? Technolegal materialization of victims' bodies and remains after the World Trade Center terrorist attacks', *Science, Technology & Human Values*, 41(4), pp. 686–708. doi: 10.1177/0162243915624145.

Toom, V. (2018a) 'Finding closure, continuing bonds, and codentification after the 9/11 attacks', *Medical Anthropology: Cross Cultural Studies in Health and Illness*, 37(4), pp. 267–279. doi: 10.1080/01459740.2017.1337118.

Toom, V. (2018b) *Cross-Border Exchange and Comparison of Forensic DNA Data in the Context of the Prüm Decision*. Brussels. Available at: www.europarl.europa.eu/RegData/etudes/STUD/2018/604971/IPOL_STU(2018)604971_EN.pdf.

Toom, V. (2020) 'Ontologically dirty knots: The production of numbers after the Srebrenica genocide', *Security Dialogue*, 51(4), pp. 358–376. doi: 10.1177/0967010620902008.

Toom, V. *et al.* (2016) 'Approaching ethical, legal and social issues of emerging forensic DNA phenotyping (FDP) technologies comprehensively: Reply to "Forensic DNA phenotyping: Predicting human appearance from crime scene material for investigative purposes" by Manfred Kayser', *Forensic Science International: Genetics*, 22, pp. e1–e4. doi: 10.1016/j.fsigen.2016.01.010.

Toom, V., Granja, R. and Ludwig, A. (2019) 'The Prüm decisions as an aspirational regime: Reviewing a decade of cross-border exchange and comparison of forensic DNA data', *Forensic Science International: Genetics*, 41, pp. 50–57. doi: 10.1016/j.fsigen.2019.03.023.

Van Oorschot, I. and M'charek, A. (2021) 'Keeping race at bay: Familial DNA research, the "Turkish community," and the pragmatics of multiple collectives in investigative practice', *BioSocieties*, 16(4), pp. 553–573.

Vidaki, A. and Kayser, M. (2017) 'From forensic epigenetics to forensic epigenomics: Broadening DNA investigative intelligence', *Genome Biology*, 18(1), p. 238. doi: 10.1186/s13059-017-1373-1.

Wienroth, M. (2018) 'Governing anticipatory technology practices: Forensic DNA phenotyping and the forensic genetics community in Europe', *New Genetics and Society*, 37(2), pp. 137–152.

Wienroth, M. (2020a) 'Value beyond scientific validity: Let's RULE (Reliability, Utility, LEgitimacy)', *Journal of Responsible Innovation*, 7(sup1), pp. 92–103. doi: 10.1080/23299460.2020.1835152.

Wienroth, M. (2020b) 'Socio-technical disagreements as ethical fora: Parabon NanoLab's forensic DNA Snapshot service at the intersection of discourses around robust science, technology validation, and commerce', *BioSocieties*, 15, pp. 28–45. https://doi.org/10.1057/s41292-018-0138-8.

Wienroth, M., Granja, R., Lipphardt, V., Nsiah Amoako, E. and McCartney, C. (2021) 'Ethics as lived practice. Anticipatory capacity and ethical decision-making in forensic genetics', *Genes,* 12(12), p. 1868.

Wienroth, M., Morling, N. and Williams, R. (2014) 'Technological innovations in forensic genetics: Social, legal and ethical aspects', *Recent Advances in DNA & Gene Sequences (Formerly Recent Patents on DNA & Gene Sequences)*, 8(2), pp. 98–103. Available at: www.ingentaconnect.com/content/ben/radnag/2014/00000008/00000002/art00006.

Williams, R. (2010) 'DNA databases and the forensic imaginary', in Hindmarsh, R. and Prainsack, B. (eds) *Genetic Suspects. Global Governance of Forensic DNA Profiling and Databasing*. Cambridge: Cambridge University Press, pp. 131–152.

Williams, R. and Johnson, P. (2004) '"Wonderment and dread": Representations of DNA in ethical disputes about forensic DNA databases', *New Genetics and Society*, 23(2), pp. 205–223.

Williams, R. and Johnson, P. (2008) *Genetic Policing. The Use of DNA in Criminal Investigations*. Cullompton, Devon: Willan Publishing.
Williams, R. and Wienroth, M. (2014a) 'Identity, mass fatality and forensic genetics', *New Genetics and Society*, 33(3), pp. 257–276.
Williams, R. and Wienroth, M. (2014b) 'Suspects, victims and others: Producing and sharing forensic genetic knowledge', *The Right to Know and the Right Not to Know*, p. 70.

PART I

DNA profiling and database governance

2

TECHNOLEGAL WORLDS IN AN ARMED CONFLICT

The forensic making of victims in Colombia*

María Fernanda Olarte-Sierra and Jaime Enrique Castro Bermúdez

Forensic world-making

Forensic knowledge for victims' identification provides testimony and information that is used and interpreted as evidence by judges to assess their decision in specific criminal cases (Cole, 2013; Rosenblatt, 2015; Fournet, 2017). Also, forensic knowledge has been put at the service of humanitarian actions in violent or war-ridden contexts, as well as in situations of disaster (whether natural or otherwise) to recover and identify victims' mortal remains and return them to their families for proper socio-cultural burial rites (Cordner and Tidball-Binz, 2017; Toom, 2018). In such situations, forensic practices of location, exhumation and identification are organised around the principle that families have the right to receive the remains of their loved ones and the right to know what may have happened to them (Bennett, 2020). In either case (that is, humanitarian or judicial uses), forensic experts are placed in a spotlight, given that their claims have direct social justice consequences that are more evident than those in conventional research science: they produce knowledge that is used in criminal and transitional justice, as well as in memory practices and humanitarian actions (Fournet, 2017; Haimes and Toom, 2014; Garibian et al., 2018; M'charek and Casartelli, 2019).

Due to its scientific basis, forensic knowledge is regarded as truth telling and has evidence status in court. It is important, however, to keep in mind that forensic experts' knowledge (like any other form of knowledge) is situated and partial. Those who produce it are placed in specific moments in history, inhabit particular geopolitical positions and can account for only the specific context in which they occur. Additionally, they draw from specific scientific understandings and modes of interpretation; they are subject to personal and technology in-built biases, operate in a specific epistemic culture (forensic science) (Cole, 2013) and are asked to interpret their perception into other epistemic cultures (the courtroom, and memory

DOI: 10.4324/9780429322358-4

practices, to mention some). Furthermore, forensic knowledge – like any other form of knowledge – has material consequences in real people's lives (Mol, 1999, 2012; M'charek, 2013; Haraway, 1988; Jasanoff, 2004; Olarte-Sierra and Castro Bermúdez, 2019; Olarte-Sierra 2022). Thus, the forensic practices of investigation, location, exhumation and identification of human remains are profoundly political activities. As Anstett and Dreyfus (2015) remark, 'the timing of exhumation always depends on the political (and sometimes geopolitical) context, such as the national politics of amnesty or the local politics of memory' (p. 5). Thus, the search, exhumation and possible identification of remains respond to an agenda of who should be looked for and when, making the incontestable decision of giving visibility to some stories of violence while others remain invisible and silenced (Ferrándiz, 2013; Korman, 2015; Garibian et al., 2018; Rousseau, 2015; Bennett, 2020). While the decision regarding when and whom to exhume is beyond the forensic experts, they are, nonetheless, the key actors conducting such activities and are those producing the knowledge.

As Guglielmucci (2017) suggests, forensic identification supposes that a legal name and a social life story make sense together, making forensic investigation crucial for identifying a person (see also Olarte-Sierra and Castro Bermúdez, 2021). Therefore, forensic experts need thorough knowledge of the socio-political, cultural and historical particularities of each case they handle to make sense of the situations they are analysing, which give the context of the possible identification. Likewise, forensic experts develop working scenarios to produce knowledge of the nature of violence to make sense of the marks it leaves on bodies (Olarte-Sierra, 2022). As a result, forensic knowledge is considered to contribute to memory practices from a privileged position, given its scientific and thus assumed objective nature (Quevedo, 2014). Nevertheless, the role of forensic knowledge for memory practices and reconciliation has been problematised. For instance, attention has been drawn to how forensics' claims about the past are problematic because they tend to disregard that their own interpretations help make sense of the past (Moon, 2013). Likewise, there are tensions and dissonances that may appear between the expectations of victims, state authorities and forensic practitioners, given that the identification of bodies can have different meanings and implications for the involved parties: e.g., some bodies are highly inconvenient for governments and public intuitions as they may evidence abuses perpetrated by the state (Crossland, 2000; Merry and Bibler-Coutin, 2014). Similarly, statements about the violent past made by forensic experts are potentially contested, depending on whether they are produced by humanitarian organisations or by state institutions, especially in cases of state violence (Moon, 2013; Collins, 2016). Hence, as Rousseau (2015) points out, forensic identification and the knowledge produced thereby must be considered in light of the socio-political and historical context in which it happens and the forensic scientist's statements be critically considered. This supposes that, despite the scientific, technical and legal elements imbued in forensic identification, it is still a social practice with clear material and tangible consequences (M'charek, 2008, 2013; Moon, 2013; Anstett and Dreyfus, 2015; Crossland, 2018).

On this matter, Williams and Wienroth (2014) remind us that identity or the act of conferring identity to a person (whether living or dead) is a political act, too, that involves social, administrative and governmental institutions. And as García-Desiter and Smith (2019) further point out, forensic identification allows bodies to become scientific material and confers to human remains a status of necro-sovereignty. This allows bodies and remains to move easily across borders (national, ideological and physical), something that the living persons who inhabited such bodies cannot always do so freely (see also M'charek and Casartelli, 2019). Therefore, as we said before, forensic identification is necessarily political. It is meant to give back personhood, judicial identity and, ideally, a cohesive social identity to those who have lost all, either due to a crime, war or another form of violence (Guglielmucci, 2017), as well as in contexts of disaster (Merli and Buck, 2015). As Toom (2016) highlights, such conferring of identity must be done in a sound technolegal framework to be legitimate and valid. Hence, knowledge produced by the forensic experts has become key for administering justice and for reconstructing the past, making ever more tangible the wider social effects of forensic identification practices (Ferrándiz, 2013; Olarte-Sierra et al., 2014; Korman, 2015; Garibian et al., 2017; Rousseau, 2015). One could further say that forensic knowledge contributes to technolegal world-building.

The ever-growing use and assumed infallibility of DNA analysis for victim identification have become political tokens too. However, despite the public trust in forensic genetics, victim identification is rare and difficult to achieve (Kruse, 2010a; Perez-Bustos et al., 2015; Bennett, 2020), and it needs much more than a DNA sample to occur. There is no quick, smooth and easy identification process (Kruse, 2010b). Instead, it has a composite materiality that results from the interrelated, interconnected and coordinated work practices of other forensic disciplines such as anthropology, medicine, dentistry and, more recently, genetics (Olarte-Sierra et al., 2014; Olarte-Sierra, 2021). The emphasis on forensic DNA as the key to solving crimes, identifying victims of mass murder and disaster and achieving victims' reparation (including the right to truth, justice and guarantees of non-repetition)[1] tends to place other forensic disciplines in a secondary position even though they are necessary for making sense of and validating DNA analysis. Furthermore, crucial socio-cultural practices of dealing with death and dead bodies are overlooked by the emphasis on DNA analysis as the main vehicle for establishing identity (Bennett, 2020). Additionally, the attention given to genetic results has also privileged the figures and numbers of identified bodies while people's stories of their suffering and violent death have been moved to the background (Olarte-Sierra, 2021).

In this chapter, we look at how forensic practices must be seen and understood as coordinated technolegal practices that sustain particular technolegal worlds in the context of the Colombian armed conflict that enact specific kinds of victims. We argue that to contribute to richer forms of victim reparation, inclusive memory practices and transparent (transitional) justice processes, forensic practices can be seen as coordinated practices which help in restoring victims' identity and personhood through the combination of technoscientific and legal regimes (Toom,

2016). Such a way to attend to forensic identification work may have the potential to contribute to the enactment of a particular history of violence that can enrich wider understandings and nuances of the Colombian armed conflict. The kind of forensic knowledge that we engage with is the one that belongs to and is produced in the framework of the criminal justice system.

Of bodies in an armed conflict

What is known as the Colombian armed conflict is considered to have started in the mid-1950s. It has been multi-faceted with varied and ever-changing actors (Krystalli, 2019). Also, it is the longest armed conflict in the region (Rettberg and Ortíz-Riomalo, 2016), with various forms of violence including murder, torture, kidnapping and enforced disappearance. The main armed actors include drug traffickers and drug lords, political and narcoguerrillas, paramilitary forces, the state and the army. All these actors have committed open violence against civilians and have affected rural and urban populations (Grupo de Memoria Histórica, 2013; Fajardo, 2014; Rodríguez Morales, 2016). The result of this has been a geography of violence that extends across the country, producing multiple groups of victims that increased rapidly with time (Centro Nacional de Memoria Histórica, 2012; Olarte-Sierra and Castro Bermúdez, 2019). According to the Unit for the Attention of Victims, a government agency, as of 1 March 2021, the Colombian armed conflict had produced more than 9,100,000 victims (Unidad para las Víctimas, 2021). However overwhelming, this number of victims is far from homogeneous. The category is multiple and complex (Guglielmucci, 2017), especially for those who seek the state's recognition and further compensation under law 1448/2011, which we will address shortly. As Krystalli (2019) points out, in Colombia, not all victims are created equal, and victimhood is subjected to hierarchies that differentiate victim groups. To access state recognition, those who consider themselves victims must face a bureaucratic process designed to confirm their history of suffering; this process not only produces legitimate victims in the eyes of the state, but such victims also correspond to the state's interests and policies (Mora-Gámez, 2016). Hence, suffering acts of violence does not automatically make a person a legitimate and official victim of the Colombian armed conflict. Being left out of the official universe of victims has implications as those so excluded cannot access 'reparation in its individual, collective, material, and symbolic forms including access to compensation for harms suffered. . . . Additional rights include preferential access to education, health, and a range of social benefits' (Krystalli, 2019: 7). It is the task of the attorney general's office to contribute to the verification of the statements of alleged victims through its different units (Mora-Gámez, 2016) and to accompany said investigations with the work of forensic experts to identify the remains of people who died.

In this context, marked with an ongoing armed conflict and multiple sorts of victims, forensic experts in the attorney general's office have been placed centre stage. Their visibility, however, has not happened in a vacuum. Legal frameworks enabled their placement and allowed for their preponderant role in the armed conflict. One of the most relevant ones was the creation of forensic teams as part of the attorney general's office's investigation units in 1994.[2] Since then, these interdisciplinary forensic groups – made up of forensic anthropologists, dentists and medical examiners – have been in charge of the location, exhumation and identification of victims of different forms of violence.[3] Also, these experts are potential expert witnesses in court, as is generally the case for forensic specialists who belong to the judicial system (Cole, 2013; Rosenblatt, 2015).

During the years of armed conflict and as a response to it, there have been numerous legislative acts that attempt either to achieve peace or to acknowledge and compensate the victims. Each of these legislative acts brought about legal and judicial frames that confer increasing weight on forensic experts and their practice. In this chapter, we focus on two partly (inter)connected laws. One is the Law of Justice and Peace enacted in 2005 (Law, 975/2005). This law is important for three reasons. First, it was a peace agreement with the paramilitary organisation known as Autodefensas Unidas de Colombia [United Self-Defence of Colombia], and the law was the transitional justice framework for the demobilised members. It involved a considerable reduction of prison sentences in exchange for providing intelligence on – among other things – the whereabouts of the bodies of people who the paramilitary disappeared, murdered and buried. The law was meant to contribute to repairing the injustices done to killed victims and their families by giving names back to those bodies and subsequently giving the bodies back to their kin. Second, the Law of Justice and Peace also helped position forensic science as crucial to judicial procedures as it mandated that forensic experts from the attorney general's office oversaw exhuming and identifying such bodies. Additionally, forensic experts were asked to present evidence in court that either confirmed or challenged the testimony of the demobilised members of the paramilitary. Third, while the legislative intentions could be considered beneficial for ending the actions of a particularly bloody armed actor, the Law of Justice and Peace was met with critiques from varied civil social sectors. Their concern was about how the law benefitted the paramilitary's versions of events while victims' testimonies remained disregarded and even ignored (Crettol and La Rosa, 2006; Uprimny et al., 2006; Díaz, 2009; Jaramillo-Marín, 2010).

The second law we attend to is the Law of Victims and Land Restitution (Law 1448 of 2011). This law recognises all victims of the Colombian armed conflict since 1985. Under this law, any person who can demonstrate that he, she or their family was subjected to violence in the context of the armed conflict can benefit from the law and receive the stipulated compensations (both symbolic and material). The law enables victims to make their petitions and complaints more visible. However, as we mentioned before, it is not a straightforward process. The Law of Victims and Land Restitution is articulated to the legislation that mandated the forensic experts of the attorney general's office to locate, exhume and identify the remains of the victims of forced disappearance and to coordinate the genetic database of missing persons.[4] It is important to mention that such legislative acts are not the sole legislation that deals with the aftermath of the armed conflict; however, the complex legal framework presented here shows that, amongst other things, it serves to organise forensic practice at the service of the judicial system. Consequently, it has technolegal materialisations 'through which human remains are forensically identified and implicated in legal regimes' (Toom, 2016, p. 689).

In the next section, through the use of two examples, we explore two technolegal worlds of victim identification that help restore victims' identity and personhood performed by the attorney general's office's forensic experts. The two cases we present here help show how forensic methods and worlds are irreducible to one discipline; all contribute to each other, have equivalent relevance and create a technolegal world where the whole is greater and more complex than the sum of its parts. The first case refers to a person disappeared by the paramilitary group AUC and thus is addressed by the Law of Justice and Peace. The other example is a youngster wrongly killed by the army. Given the conditions of his disappearance and death, his case can be considered and falls under the Law of Victims and Land Restitution.

We attend to the identification practices carried out by forensic experts in Colombia's attorney general's office. These experts work amid an ongoing armed conflict. Our analysis stems from two anthropological branches. One is the anthropology of science, from which we understand that scientific products (such as the production of knowledge) are socio-cultural (M'charek, 2013; Law and Moser, 2012; Law, 2015; Blume, 2017). The other is social anthropology, which enables us to address the facts, practices and social objects of the armed conflict as contextualised matters (Uribe, 2008; Castillejo-Cuellar, 2014). We, the authors, want to mention that Olarte-Sierra is an anthropologist of science interested in the field of forensic practices, and Castro Bemúdez is a seasoned forensic expert of the AGO with more than 23 years of experience. Since 2016, we have embarked on a fruitful journey of collaborative research and publication with the intention of conceptualising our nourished discussions. One of the topics we have addressed during our research was the value and role of DNA analysis vis à vis other forensic practices.

On people and figures

As said before, we propose that forensic identification is a composite practice made of diverse disciplines that need to collaborate with one another to identify a person. In doing so, these practices must also account for the person's story of violence (Guglielmucci, 2017; Bennett, 2020). In contexts of transitional justice, this knowledge is also put at the disposition of practices aiming at the administration of justice, the establishment the truth and the reparation of victims (Fournet, 2017). Here, we articulate the conceptual richness and political nature of forensic practices of investigation and identification as a composite by briefly discussing two exemplary cases. The first example is Leidy's case. In it, we show how the centrality of DNA testing contributes to blurring the victim's identity and life history while serving the purpose of returning to a family the remains of their missing child. Nonetheless, we advance that the emphasis on DNA alone considers the bodies produced by the conflict while it makes unknowable (Smith and Garcia-Deister,

2017) the reasons specific populations faced particular kinds of violence (CNMH, 2015; Olarte-Sierra, 2022). In the second example, the case of Alejandro, we show how the comprehensive forensic investigation of his death and body turned this youngster into a very specific victim of the Colombian armed conflict. Through these two examples, we show how differentiated victims' technolegal worlds are represented: that is, worlds that articulate and enable specific kinds of victimhood in the Colombian armed conflict.

Data or stories?

Leidy[5] was a transgender sex worker who, in early 2000, was disappeared by the former paramilitary group Autodefensas Unidas de Colombia (AUC) [United Self-Defence of Colombia]. One of the men involved in her disappearance (and assassination) provided information on the whereabouts of her remains. This was part of the intelligence he provided to receive a reduced prison sentence in the framework of Law of Justice and Peace. Leidy's body was recovered where the *postulado*[6] indicated (Olarte-Sierra, 2022). As per protocol, Leidy's body was sent to one of the identification laboratories that the attorney general's office has across the country. Once there, a forensic anthropologist cleaned and individualised her (i.e., divided her bones from other bones recovered at the gravesite) and proceeded with the anthropological examination. The result of this first approach to Leidy's remains was that her body was classified as male. To confirm this classification, the forensic anthropologist sent a bone to the genetics laboratory. The DNA analysis confirmed that the bone belonged to a man, Pedro, who was reported missing by his parents some years ago. After such a confirmation, the fact that Leidy was living as a female transgender sex worker before her death became secondary. The confirmed legal name, Pedro, was the one registered as a victim in the long list of conflict victims missing and located by the attorney general's office in the context of the Law of Justice and Peace. Pedro's parents received the remains of their son, but Leidy's story, like the stories of many other transgender people who lost their lives in the armed conflict due to their gender identity, was made invisible.[7]

Here we have a case of identification. A successful story of giving a name to a collection of human remains. But was it truly an act of returning identity and personhood to the person who was killed and disappeared? The forensic DNA analysis identified a missing man, and on the face of it, a name, and delivering his remains to his relatives was possible. Thus, Leidy's name was added to the 'solved' cases while the reason she was killed (i.e., being a transexual sex worker, as stated by the *postulado*) was left behind and made unknowable (Olarte-Sierra, 2022). The technolegal world produced here is a combination of the work of forensic identification (that started with the physical anthropological assessment of Leidy's body and ended with DNA analysis that matched the genetic information of a couple searching for their missing son) and the legal imposition of pronouncing someone's death only by his or her legal name. Additionally, this technolegal world allows for disregarding the details of a *postulado*'s testimony, despite the fact that such testimonies constitute the very basis of the law and the administration of transitional justice. Consequently, it is a world in which the identification of human remains is possible and *postulados* receives the benefits of Law 975/2005, yet the life history of the person on whom the violence was perpetrated is made invisible, even though that very story is what helps make sense of her untimely death and the doings of the paramilitary regarding lesbian, gay, transgender and queer (LGBTQ) populations from a forensic point of view.

Positives and false positives

In September 2008, the media uncovered the case of what is known in Colombia as 'false positives'; the correct name of the crime is extrajudicial executions: assassination of unarmed civilians by army members. The news told the story of several young and underprivileged men coming from the city of Soacha and the south of Bogota (the capital of Colombia). They had been reported missing by their relatives and had been found dead later. The heart of the story was that these youngsters were presented as guerrilla casualties in combat with the Colombian Armed Forces while their families (mainly mothers) claimed that they were innocent youngsters lured by strangers with false job promises. It soon became clear that what had happened in Soacha and southern Bogota were not isolated events. Across the country, young men were lured away from their homes to be kidnapped, disappeared, killed and then presented as guerrilla casualties (Pachón, 2009; Londoño Carvajal, 2011; El Nuevo Siglo, 2012; Palau, 2020). After months of work, investigators in the attorney general's office were able to establish that the youngsters were, in fact, innocent citizens who had been killed by army members in response to and in the framework of incentives to stimulate the war against terrorism. The incentives supposed that with each capture (dead or alive), the military received substantial benefits in the form of promotions, holidays and shared monetary rewards (Pachón, 2009; Londoño Carvajal, 2011; Cárdenas and Villa, 2013; Salamanca, 2014; Borbón Torres, 2019). The phenomenon of these extrajudicial executions was an effect of the Policy of Democratic Security and Defence launched in 2003 by former President Alvaro Uribe Velez (Ministerio de Defensa, 2003; Galindo 2009; Londoño Carvajal, 2011; Cárdenas and Villa, 2013), and it is considered a crime against humanity in humanitarian international law.[8]

As said before, forensic identification supposes the establishment of identity in terms not only of the person's name but also of their life history. To decide whether or not a person is a victim of a case of extrajudicial execution and falsely labelled as guerrilla member, forensic experts need much more than the person's name, which can be established through a forensic DNA analysis. They must determine that the victim does not belong to a guerrilla group and then build their case. This is achieved by signalling possible discrepancies between the circumstances in which the death took place (as reported either by his or her killers or in post-mortem records), and the story of the person who was killed as told by those close to him or her. In this particular case, forensic experts needed to overlap the story of the presumed guerrilla member to the one of an innocent citizen and then establish what does not make sense in either of the stories (Olarte-Sierra and Castro Bermúdez, 2021).[9] Hence, the practice of forensic investigation and the establishment of specific kinds of victimhood include working directly on the person's body *and* the reconstruction of the person's life history. As we will show briefly, these forensic practices of investigation and identification working with and in a specific legal framework are part of a technolegal world that enacts the victims of extrajudicial executions.

Amongst the elements that contradicted the official versions of the casualties reported by the army was the poor staging of the deaths of the claimed guerrilla members. The death scenes included bodies wearing brand new and nameless guerrilla uniforms, new and mixed-up wellington boots (i.e., the right foot wearing the left boot) and the employment of armament registered as confiscated by the army years before (Pachón, 2009; Semana, 2008, 2010). In their investigation of such deaths, forensic experts paired the bodies' ill staging and death reports made by army members with the accounts of the youngsters' relatives and witnesses, and in those proceedings, the stories of the so-called casualties became incongruent and contradictory.

This was the case of Alejandro, a young baker who was reported missing by his sister, found dead months later and buried, and labelled by the army as a guerrilla casualty. Co-author Jaime Enrique Castro Bermúdez was the forensic investigator in this case, and he explains:

> I saw the photos of the case and I could see errors in the [death] scene, in the body. He had a gun in his right hand and that, that could not be right. I had already spoken with the family and the sister told me that he was not a guerrilla member, that he was a baker. . . . But what is more, she told me that as a child he had suffered an accident, that his right arm had a permanent injury and it did not work properly, it didn't develop as it should. And of course, when examining the body there it was, the difference in muscle tone and in the development of the right arm compared to the left. He could not have fired with that hand, that arm had no muscle tone and therefore he could not hold the weapon they had put on him.
>
> *(Conversation between authors, November, 2017, our translation)*

Due to Jaime and his colleagues' investigation of Alejandro's story, a more comprehensible and complete return of his personhood was possible. The determination of his name was not the final point of forensic practice. Rather, he was given an identity, a history, a family and a profession, all of which were needed to make him into an innocent citizen and thus a legitimate victim of a case of 'false positives'. It was only through the combined efforts of forensic practices of varied disciplines that his innocence could be established and his victimhood demonstrated. In terms of justice, those responsible for Alejandro's death faced prosecution and his family received reparations by receiving his remains and by having his name cleared, as well as the benefits they were entitled to by the Law of Victims and

Land Restitution. In terms of the wider social effect of forensic practices of identification, forensic knowledge enriched the social accounts of 'false positives' as an undeniable atrocity provoked by the state towards its own citizens (see Olarte-Sierra and Castro Bermúdez, 2021). The technolegal world in which Alejandro and the other approximately 6,400 victims (JEP, 2021) are enacted is one made not only by domestic legislation (Law of Victims and Land Restitution) but also by international humanitarian law and technoscientific practices aimed at debunking army members' statements of winning the war against terrorism. These technolegal devices allowed the shifting of these youngsters' identities from guerrilla members to victims of extrajudicial executions.

Coordinated forensic practices

In this chapter, we have argued that forensic practices are composites of disciplines that must work together in order to achieve identification. Furthermore, interdisciplinary forensic work has the potential to come closer to victims' reparation, memory practices and justice processes when seen and understood as technolegally (Toom, 2016) coordinated practices. As a result, forensic work can contribute to the enactment of a complex history of violence and thus enrich the understanding of the armed conflict while co-producing an explanation of its nuances since forensic practices, like other scientific practices, have material effects that go beyond the limits of the judicial and criminal systems they serve (Olarte-Sierra and Castro Bermúdez, 2019; Olarte-Sierra, 2022).

We presented two technolegal worlds that represent particular kinds of victims. We focused our attention on the long-lasting and still active Colombian armed conflict that has produced more than nine million victims (Unidad para las Víctimas, 2021) by multiple and ever-changing changing actors. We looked at various legal and judicial acts that provide forensic experts in the attorney general's office a central role in the conflict while being in charge of exhuming and identifying the victims. To develop our argument, we addressed two cases of identification. The two cases we presented here signal the potential that forensic experts and practices have for accounting for and co-producing the Colombian armed conflict and make evident the possible consequences not only for the victims and their relatives but also in terms of justice and memory practices. In this sense, as Guglielmucci (2017) points out, an exhaustive identification is not solely the judicial and administrative act of identifying a body with a name but also the social act of returning a cohesive social identity to the person who died (see also Toom, 2018). We add that such a fuller identification is only achievable when forensic experts of the varied disciplines work in a combined effort to reach a comprehensive identification.

In the case of Leidy – who is part of one technolegal word – a well-known history of abuse and gender violence towards an entire population is made invisible and unknowable. By deciding to prioritise a victim's name and match in the attorney general's office genetic databases, the negative consequences on transitional justice, victims' reparation and historic memory are undeniable (Olarte-Sierra et al.,

2014; Olarte-Sierra, 2022). The victim's legal name (Pedro, not Leidy) was relevant for national figures of identified bodies and victims of the armed conflict. However, her killers did not face charges of gender violence since that was not made visible in the official investigation. That is, what was important for the 975/05 law was to deliver remains, not to tell the stories behind the selective deaths (Olarte-Sierra, 2022). For Alejandro, who is part of another technolegal world, the combined efforts of diverse forensic disciplines enacted a fuller and richer account of the violence perpetrated on him. The reconstruction of the events, the particularities of his own history (e.g., his accident when he was a child) and the careful study of the death scene and subsequent documentation produced by the army allowed forensic experts not only to provide a name and to return him to his relatives but also to present a fully fleshed-out story that contributed to the social understanding and the judicial intervention of the despicable phenomenon of the 'false positives' (Olarte-Sierra and Castro Bermúdez, 2021).

Notes

* This project has received funding from the European Union's Horizon 2020 research and innovation programme under the Marie Skłodowska-Curie Grant Agreement No. 898537

1 Reparation is form of delivering justice to victims of human rights violations in transitional justice scenarios. They include but are not limited to compensation and restitution and must also include the right to truth, justice and guarantees of non-repetition. See UNHCHR, 2008.

2 The Colombian Political Constitution changed in 1991. Amongst the institutions it ordered created was the attorney general's office, a state institution in charge of administering justice to civilians who commit crimes.

3 Forensic experts of the National Institute of Legal Medicine and Forensic Science support identification procedures, but only rarely and after the expressed petition of the AGO can they preform exhumations

4 It is important to emphasise that the practice of forensic genetics since the early 2000s precedes this approach and legislative event. Its origin was for paternity testing performed to allocate responsibility to fathers who did not take financial care of their children (Olarte-Sierra and Diaz del Castillo, 2014). Before Law 1408/10, genetic tests were practiced to establish the identity of disappeared victims whose bone remains were recovered. Therefore, this new legislation fits into a well-known and accepted – both socially and legally – scientific proceeding.

5 All names are pseudonym. In Leidy's case, however, we kept the Anglicism of the original name usually written in Colombian Spanish.

6 This is the name given to the paramilitary members who decided to seek justice under Law of Justice and Peace.

7 For a thorough analysis of Leidy's case, see Olarte-Sierra, 2022.

8 Sadly, extrajudicial executions are taking place again in Colombia, and new forms of paramilitary actions are occurring across the national territory (Conexión Capital, 2020). The AGO has been working on these cases since 2005 (Olarte-Sierra and Castro Bermúdez, 2019), and currently, the Special Jurisdiction for Peace (as mentioned before, an institutional product of the peace agreement with the FARC-EP) is conducting investigations and administering transitional justice to those army members who have confessed to the crime. Nonetheless, given the nature of this crime and the fact that the armed forces are responsible for these deaths, since 2004, the Office of the Prosecutor of the International Criminal Court is advancing a preliminary investigation for crimes against humanity (ReD, 2020).

9 In another paper, we develop this argument in full. See Olarte-Sierra and Castro Bermúdez, 2021.

References

Anstett, Elizabeth, y Jean-Marc Dreyfus. 2015. *Human Remains and Identification: Mass Violence, Genocide, and the "forensic Turn"*. Manchester: Manchester University Press.

Bennett, Caroline. 2020. "Is DNA always the answer?" In Roberto C. Parra, Sara C. Zapico, and Douglas H. Ubelaker (eds.) *Forensic science and humanitarian action: interacting with the dead and the living*, 521–534. New York: John Wiley & Sons.

Blume, Stuart. 2017. *Immunization: how vaccines became controversial*. London: Reaktion Books.

Borbón Torres, Julián Alfredo. 2019. "La política de defensa y seguridad democrática en el Estado social de derecho". *Revista IUSTA* 1 (50): 97–120. https://doi.org/10.15332/1900-0448.2019.0050.04.

Cárdenas, Ernesto, y Edgar. Villa. 2013. "La Política de Seguridad Democrática y las Ejecuciones Extrajudiciales [The Democratic Security Policy and Extrajudicial Executions]". *Ensayos Sobre Política Económica* 31: 64–72.

Castillejo-Cuellar, Alejandro. 2014. "La localización del daño. Etnografía, espacio y confesión en el escenario transicional colombiano". *Horizontes Antropológicos* 20: 213–236.

Centro Nacional de Memoria Histórica. 2012. *Estadísticas Del Conflicto Armado En Colombia. Official Site*. Bogotá: Centro Nacional de Memoria Histórica www.centrodememoriahistorica.gov.co/micrositios/informeGeneral/estadisticas.html.

Centro Nacional de Memoria Histórica. CNMH. 2015. *Aniquilar la diferencia. Lesbianas, Gays, Bisexuales y Transgeneristas en el marco del conflicto armado colombiano [Annihilate the difference. Lesbians, Gays, Bisexuals]*. Bogotá: Centro Nacional de Memoria Histórica.

Cole, Simon A. 2013. "Forensic Culture as Epistemic Culture: The Sociology of Forensic Science". *Studies in History and Philosophy of Biological and Biomedical Sciences* 44 (2013): 36–46.

Collins, Cath 2016. *Respuestas Estatales a La Desaparición Forzada En Chile: Aspectos Forenses, Policiales Y Jurídicos*. Chile: Observatorio de Justicia Transicional. Universidad Diego Portales.

Conexión Capital. 2020. "Denuncian 20 nuevos casos de falsos positivos en Arauca". *Conexión Capital*, 26 de octubre de 2020. https://conexioncapital.co/denuncian-20-nuevos-casos-de-falsos-positivos-en-arauca/.

Congreso de Colombia. 2005. Law 975/2005.

Congreso de Colombia. 2010. Law 1408/2010.

Congreso de Colombia. 2011. Law 1448/2010.

Cordner, Stephen, and Morris Tidball-Binz. 2017. "Humanitarian Forensic Action—Its Origins and Future". *Forensic Science International*. https://doi.org/10.1016/j.forsciint.2017.08.011.

Crettol, Monique, and Anne-Marie La Rosa. 2006. "The Missing and Transitional Justice: The Right to Know and the Fight Against Impunity". *International Review of the Red Cross* 88 (862): 1–8.

Crossland, Zoë. 2000. "Buried Lives: Forensic Archaeology and the Disappeared in Argentina". *Archaeological Dialogues* 7 (2): 146–519.

Crossland, Zoë. 2018. "Forensic Afterlives". *Signs and Society* 6 (3): 622–647. https://doi.org/10.1086/699597.

Díaz, Ivonne Leadith. 2009. "El rostro de los invisibles víctimas y su derecho a la verdad, justicia, reparación y no repetición". In David Bondia García and Manuel Ramiro Muñoz (eds.) *Víctimas invisibles, conflicto armado y resistencia civil en Colombia*, vol. 1, 37–68. Madrid: Huygens Editorial.

El Nuevo Siglo. 2012. "Falsos positivos: víctimas no recogían café". *Redacción*, 8 de octubre de 2012. www.elnuevosiglo.com.co/articulos/10-2012-falsos-positivos-victimas-no-recogian-cafe.

Fajardo, Dario. 2014. *Estudio sobre los orígenes del conflicto social armado, razones de su persistencia y sus efectos más profundos en la sociedad colombiana*. Bogotá: Centro Nacional de Memoria Histórica.

Ferrándiz, Francisco. 2013. "Exhuming the Defeated: Civil War Mass Graves in 21st-Century Spain". *American Ethnologist* 40 (1): 38–54.

Fournet, Caroline. 2017. "Forensic truth? Scientific evidence in international criminal justice". *Humanity Journal,* 12 July. http://humanityjournal.org/blog/forensic-truth/

Galindo, Carolina 2009. "De la Seguridad Nacional a la Seguridad Democrática: nuevos problemas, viejos esquemas". En Carolina Galindo y Adolfo Chaparro (eds.) *Génesis y transformaciones del estado nación en Colombia: una mirada topológica a los estudios sociales desde la filosofía*, 219–250. Bogotá: Editorial Universidad del Rosario.

García-Desiter, Vivette, and Lindsay Smith. 2019. "Migrant Flows and Necro-Sovereignty: The Itineraries of Bodies, Samples, and Data Across the US-Mexico Borderlands". *Biosocieties*: 1–18.

Garibian, Sévane, Élisabeth nstett, y Jean-Marc Dreyfus, editores. 2017. *Restos humanos e identificación. Violencia de masa, genocidio y el "giro forense"*. Buenos Aires: Miño y Dávila Editores.

Grupo de Memoria Histórica. 2013. *¡Basta YA! Colombia: memorias de guerra y dignidad*. Bogota.

Guglielmucci, Ana. 2017 "Identidades fragmentadas: los procesos de identificación forense en casos de desaparición forzada". *Revista de Antropología* 30: 105–136.

Haimes, Erica, and Victor Toom. 2014. "Hidden in full sight: Kinship, science and the law in the aftermath of the Srebrenica genocide". *New Genetics and Society*33 (3): 277–294.

Haraway, Donna. 1988. "Situated Knowledges: The Science Question in Feminism and the Privilege of Partial Perspective". *Feminist Studies* 14 (3): 575–599.

Jaramillo-Marín, Jefferson. 2010. "Reflexiones sobre los 'usos' y 'abusos' de la verdad, la justicia y la reparación en el proceso de Justicia y Paz colombiano (2005–2010) [Reflections on the 'uses' and 'abuses' of truth, justice and reparation in the Colombian Justice and Peace process (2005–2010)]". *Papel Político* 15 (1): 13–46.

Jasanoff, Sheila. 2004. "Ordering Knowledge, Ordering Society". In Sheila Jasanoff (ed.) *States of Knowledge. The Co-Production of Science and the Social Order*, 13–45. Routledge: Oxfordshire.

JEP—Jurisdicción Espacial para la Paz. 2021. Caso 03. Asesinatos y desapariciones forzadas presentados como bajas en combate por agentes del Estado. https://www.jep.gov.co/especiales1/macrocasos/03.html

Korman, Rémy. 2015. "Bury or display? The politics of exhumation in post-genocide Rwanda". In Elizabeth Anstett y Jean-Marc Dreyfus (eds.) *Human Remains and Identification: Mass Violence, Genocide, and the "forensic Turn"*. Manchester: Manchester University Press.

Kruse, Corinna. 2010a. "Producing Absolute Truth: CSI Science as Wishful Thinking". *American Anthropologist* 112 (1): 79–91. DOI: 10.1111/j.1548–1433.2009.01198.x

Kruse, Corinna. 2010b. "Forensic Evidence: Materializing Bodies, Materializing Crimes". *European Journal of Women's Studies* 17 (3): 363–377. https://doi.org/10.1177/1350506810377699.

Krystalli, R. 2019. "We Are Not Good Victims: Hierarchies of Suffering and the Politics of Victimhood in Colombia". Doctoral Thesis, The Fletcher School of Law and Diplomacy. Medford.

Law, John. 2015. "STS as Method". In U. Felt, R. Fouché, C. Miller, and L. Smith-Doerr (eds.), *The Handbook of Science and Technology Studies*, 31–58. Cambridge: MIT Press.

Law, John, y Ingunn Moser. 2012. "Contexts and Culling". *Science, Technology, & Human Values* 37 (4): 332–354. https://doi.org/10.1177/0162243911425055.

Londoño Carvajal, Germán. 2011. "Estado seguro—ciudadano amenazado: El paradigma de la seguridad y sus gestos simbólicos". *Revista Internacional de Filosofía* 4: 119–129.

M'charek, Amade. 2008. "Silent Witness, Articulate Collective: DNA Evidence and the Inference of Visible Traits". *Bioethics* 22 (9): 519–528.

M'charek, Amade. 2013. "Beyond Fact or Fiction: On the Materiality of Race in Practice". *Cultural Anthropology* 28 (3): 420–442.

M'charek Amade, and Sara Casartelli. 2019. "Identifying Dead Migrants: Forensic Care Work and Relational Citizenship". *Citizenship Studies*. Published ahead of print: https://doi.org/10.1080/13621025.2019.1651102

Merli, Claudia, and Trudi Buck. 2015. "Forensic identification and identity politics in 2004 post-tsunami Thailand: Negotiating dissolving boundaries". *Human Remains and Violence: An Interdisciplinary Journal* 1 (1): 3–22.

Merry, Sally Engle, and Susan Bibler Coutin. 2014. "Technologies of Truth in the Anthropology of Conflict: AES/APLA Presidential Address, 2013". *American Ethnologist* 41 (1) (February 16, 2014): 1–16.

Ministerio de Defensa. 2003. *Política de Defensa y Seguridad Democrática*. República de Colombia.

Mol, Annemarie. 1999. "Ontological Politics. A Word and Some Questions". *The Sociological Review* 47 (1_suppl) (May 1, 1999): 74–89. https://doi.org/10.1111/j.1467-954X.1999.tb03483.x.

Mol, Annemarie. 2012. "Mind Your Plate! The Ontonorms of Dutch Dieting". *Social Studies of Science* 43 (3): 379–396. https://doi.org/10.1177/0306312712456948.

Moon, Claire. 2013. "Interpreters of the Dead: Forensic Knowledge, Human Remains and the Politics of the Past". *Social & Legal Studies* 22 (2): 149–169.

Mora-Gámez, Fredy. 2016. "Reconocimiento de víctimas del conflicto armado en Colombia: Sobre tecnologías de representación y configuraciones de Estado". *Universitas Humanistica* 82: 75–101.

Olarte-Sierra, Maria Fernanda. 2022. "(Un)Doing the Colombian armed conflict: Forensic knowledge, contradicting bodies, unsettling stories". *Social Anthropology* 30 (3): 19–37.

Olarte-Sierra, Maria Fernanda, and Jaime Enrique Castro Bermúdez. 2019. "Notas forenses: conocimiento que materializa a los cuerpos del enemigo en fosas paramilitares y falsos positivos [Forensic Notes: Knowledge That Materializes the Enemy's Bodies in Paramilitary Pits and False Positives]". *Antípoda Revista de Antropología y Arqueología* 34: 119–140.

Olarte-Sierra, Maria Fernanda, and Jaime Enrique Castro Bermúdez. 2022. "De guerrilleros a víctimas; de héroes a perpetradores: movimientos y relaciones del conflicto armado colombiano y la práctica de identificación forense en el caso de los Falsos Positivos". *Papeles del CEIC* 2: 1–17.

Olarte-Sierra, Maria Fernanda, Adriana Diaz del Castillo, Natalia Pulido Ronchaquira, Nathalia Cabrera Villota, y Roberto Suarez Montañez. 2014. "Verdad e incertidumbre en el marco del conflicto en Colombia: una mirada a los sistemas de información como prácticas de memoria". *Universitas Humanística* 79: 233–254.

Pachón, Mónica. 2009. "Colombia 2008: éxitos, peligros y desaciertos de la política de seguridad democrática de la administración Uribe". *Revista de Ciencia Política* 29 (2): 327–353. http://dx.doi.org/10.4067/S0718-090X2009000200005.

Palau, Mariana. 2020. "The 'False Positives' Scandal That Felled Colombia's Military Hero". *The Guardian*, 19 November 2020.

Perez-Bustos, Tania, Maria Fernanda Olarte-Sierra, and Adriana Diaz del Castillo. 2015 "Working with Care: Experiences of Invisible Women Scientists Practicing Forensic Genetics in Colombia". En Eden Medina, Ivan Marques, y Christina Holmes (editado por) *Beyond Imported Magic*, 67–86. Cambridge, MA: MIT Press.

Quevedo, Helka. 2014. *Textos corporales de la verdad. memoria histórica y antropología forense*. Bogotá: Centro Nacional de Memoria Histórica.

ReD-Rodeemos el Dialogo. 2020. *Caso 003: Muertes ilegítimamente presentadas como bajas en combate por agentes del Estado o «falsos positivos»*. Rodeemos el Diálogo. https://rodeemoseldialogo.org/caso-003-muertes-ilegitimamente-*presentadas-como-bajas-en-combate-por-agentes-del-estado-o-falsos-positivos/*.

Rettberg, Angelika, and Juan Felipe Ortíz Riomalo. 2016. "Golden Opportunity, or a New Twist on the Resource—Conflict Relationship: Links Between the Drug Trade and Illegal Gold Mining in Colombia". *World Development* 84: 82–96.

Rodríguez Morales, Tania Gabriela. 2016. "Geografía del terrorismo en Colombia: Una visión retrospectiva". *Revista de Paz y Conflictos* 9 (2): 179–198.

Rosenblatt, Adam. 2015. *Digging for the Disappeared: Forensic Science After Atrocity*. Stanford, CA: Stanford University Press.

Rousseau, Nicky. 2015. "Identification, Politics, Disciplines: Missing Persons and Colonial Skeletons in South Africa". In Elizabeth Anstett and Jean-Marc Dreyfus (eds.) *Human Remains and Identification: Mass Violence, Genocide, and the "Forensic Turn*, 175–202. Manchester: Manchester University Press.

Salamanca, Claudia. 2014. "Cuatro Imágenes de Falsos Positivos". presentado en Historia del Arte y Poder. Fase 3: Historias de Investigación, Museo Nacional de Colombia, septiembre 15. www.youtube.com/watch?v=APKUdB9Ieto&t=639s.

Semana. 2008. "Ya son 46 los jóvenes desaparecidos que fueron reportados como muertos en combate". *Semana*, 26 de septiembre de 2008. www.semana.com/nacion/conflicto-armado/articulo/ya-46-jovenes-desaparecidos-fueron-reportados-como-muertos-combate/95578-3.

Semana. 2010. "Los casos olvidados de los 'falsos positivos'". *Semana*, 17 de julio de 2010. www.semana.com/nacion/articulo/los-casos-olvidados-falsos-positivos/119416-3.

Smith, L. A., and V. García-Deister. 2017. "Capturing Los Migrantes Desaparecidos: Crisis, Unknowability, and the Making of the Missing". *Perspectives on Science* 25 (5): 680–697.

Toom, Victor. 2016. "Whose Body Is It? Technolegal Materialization of Victims' Bodies and Remains after the World Trade Center Terrorist Attacks". *Science, Technology & Human Values* 41 (4): 686–708.

Toom, Victor. 2018. "Finding Closure, Continuing Bonds, and Codentification After the 9/11 Attacks". *Medical Anthropology* 37 (4): 267–279. https://doi.org/10.1080/01459740.2017.1337118.

Unidad para las Víctimas. 2021. "Regsitro Único de Víctimas". Accessed 9 December 2020. http://rni.unidadvictimas.gov.co/RUV.

United Nations High Commissioner for Human Rights. 2008. *Rule-of-Law Tools for Post-Conflict States: Reparations Programmes*. Geneva and New York, United Nations.

UN Office of the High Commissioner for Human Rights (OHCHR), Rule of Law Tools for Post-Conflict States: Reparations Programmes, 2008, HR/PUB/08/1, available at: https://www.refworld.org/docid/47ea6ebf2.html [accessed 28 October 2022].

Uprimny, Rodrigo, Catalina Botero, Esteban Restrepo, and María Paula Saffon. 2006. *¿Justicia transicional sin transición? Reflexiones sobre verdad, justicia y reparación en Colombia [Transitional Justice without Transition? Reflections on Truth, Justice and Reparation in Colombia]*. Bogotá: Dejusticia. Centro de Estudios de Derecho, Justicia y Sociedad.

Uribe, María Victoria. 2008. "Mata, que Dios perdona. Gestos de humanización en medio de la inhumanidad que circunda a Colombia". En *Veena Das: Sujetos del dolor, agentes de dignidad*, 171–191. Bogotá: Pontificia Universidad Javeriana. Universidad Nacional de Colombia.

Williams, Robin, y Matthias Wienroth. 2014. "Identity, Mass Fatality and Forensic Genetics". *New Genetics and Society* 33 (3): 257–276. https://doi.org/10.1080/14636778.2014.946005.

3

TRAVELLING PROMISES

Forensic DNA databases in Brazil's technolegal setting

Vitor Simonis Richter and Luiza Louzada

Since 2009, after a collaboration agreement with the United States' Federal Bureau of Investigations (FBI), Brazil has increased its efforts to expand the use of forensic DNA databases in its technolegal scenery.[1] Although some states in Brazil already had their 'in-house' DNA databases, a network connecting states to a national system was not available. In 2012, a federal law regulating the use of this technology was approved after a year and a half of legal treatment in the Senate, a fast legislative process by Brazilian standards. Subsequently, the network that comprises this national database became the largest in a single country outside the United States' Combined DNA Index System (CODIS) and promised to address Brazil's urgent and dramatic public security issues through the sampling and storage of convicted individuals' and crime scene genetic profiles. Yet many problems were and still are to be dealt with, including Brazil's painful history of a military dictatorship (1964–1986) and the memories regarding that violent era.

In this chapter, we draw from our empirical research on the introduction of forensic DNA databases in Brazil to give an account on subjects that help us understand the history and practices surrounding this technology in the country. First, we address early dynamics of the introduction of the forensic DNA databases to outline some associations between technology, state, legal actors and commercial interests that were important to shaping the introduction process and the legal framing of this 'traveling technology' when it arrived in Brazil. Second, we address critical responses to the legal framing, focusing on a Supreme Court public hearing on the constitutionality of Brazilian DNA database law. This hearing brings us to the next subject concerning Brazilian's forensic DNA database's practical challenges that emerge from the relation with other institutions, infrastructures and the memory of the dictatorship's history of torture inside prisons. This part also addresses the expansion of database inclusion criteria to those convicted for a *crime doloso* (crime committed with intent). The chapter ends bringing forward

DOI: 10.4324/9780429322358-5

the role of Brazilian police organisations and infrastructure in making the country's forensic DNA database model. Our intention is to highlight that the Brazilian database's focus on mandatory sampling of incarcerated individuals and all its challenges inside Brazilian prisons is not just a choice made by biotechnology companies' lobby and forensic genetics professionals dazzled by the UK and US databases' promises. It is also a recognition by these same stakeholders that the Brazilian police model brings unavoidable difficulties for any police practice that relies on crime scene preservation.

'Pass the law and the money will come': Investing in the infrastructure's promise

Brazilian forensic experts (*peritos*) have been engaged in introducing forensic genetics since the mid-1990s. This introduction process, however, relied mostly on forensic experts' individual efforts scattered around the country's forensic biochemistry laboratories. Forensic expertise and technologies weren't part of national projects and security policies until the late1990s and early 2000s. According to several Brazilian public security experts (Misse, 2006; Soares, 2007), Brazil has never experienced a period of continued and integrated national policy for public security through the 20th century. Its public security policies are well captured by what Luiz Eduardo Soares (2000) has called a 'pendular movement', alternating between more repressive policies towards marginalized populations and, after the redemocratisation of 1985, policies of the valorization of human rights and the Constitution of 1988. As a consequence of this pendular movement, Michel Misse (2006) argues that Brazilian public security institutions never developed proper and integrated structures of information and collaboration between different police offices (*delegacias*) and other institutions, such as the forensic institutes.

During the early 2000s, however, the *Secretaria Nacional de Segurança Pública* (SENASP) [National Public Security Office], created in 1997, began to engage in a series of national plans for public security that sought to change the orientation from a 'national security doctrine' that characterized Latin America security policies throughout the 20th century to citizen-oriented policies based on the language of human rights. As highlighted by Misse (2006), SENASP's policy statements began to include concerns such as the following:

> It's urgent a radical reform in forensic sciences, involving the establishment of collaboration agreements with universities and research institutions, permanent technical updating of personnel and its valorization, rigorous recruitment and education, re-equipment and capilar diffusion of forensic services and decentralization of laboratories.
>
> *(SENASP apud Misse, 2006)*

This institutional orientation fomented the crucial step taken in order to put Brazil on track to create its own national DNA database: a collaboration agreement with the FBI in 2009 gave access to the FBI's CODIS system.[2] After the agreement was signed, Brazilian forensic experts travelled to the United States for training at FBI facilities. These experts became responsible for training their Brazilian colleagues in federal police's and states' forensic services. It also demanded a new infrastructure for the arriving technology, including new and better-equipped laboratories. According to a 2012 SENASP report,[3] only six states out of twenty-seven had working forensic genetics laboratories by 2011, the year the DNA database bill was submitted to Congress. Fifteen other states were able to conduct DNA analysis, but only seven had a system to register the chain of custody.

With the prospect of the DNA database, forensic science began to play a more prominent role in Brazilian public security policies.[4] It was accompanied by and heavily dependent on forensic genetics, its laboratory infrastructure and machinery and the sociotechnical imaginary[5] (Jasanoff, 2015) surrounding DNA and biotechnology. The combination of these elements also relied on what Robin Williams (2010) has called 'forensic imaginary'. This notion combines two principles. First, it's the assumption that the individuation of any object is always possible. When this principle is challenged, the faith of technological and scientific development is foregrounded, with the argument that while it is not possible yet, identification will be within reach in the near future. The second principle is based on the assumption that exchange of matter or vital material happens every time two individuals make physical contact or an individual is present in a physical space (M'charek, 2008).

The 'forensic imaginary' that enacts promising images of using science and technology in police investigations became strategic to the whole Brazilian forensic sciences pledges for more funding, training and protagonism towards forensic experts in formulating public security policies. It became an icon of investment in 'modern' and 'advanced' technologies that promised to face one of the nation's most urgent problems, the high crime rate and low rates of crime solving. According to the *Instituto de Pesquisas Econômicas Aplicadas* (IPEA) [Applied Economics Research Institute],[6] in 2017, Brazil registered 65,602 homicides.[7] Alongside this information, the 2019 IPEA's report *Atlas da Violência* [Atlas of Violence] says that it's not possible to measure the elucidation of the crime rate because several states don't keep these records. IPEA's report estimates that only around 10% to 20% of homicides are solved in Brazil.

The combination of genetic expertise, the precision of biotechnology and the agility of databases was made the holder of an appealing image of the future. A future in which public security agents such as forensic scientists, and not just the police officers often seen as violent, corrupt and with less education (school years), would become more efficient in presenting the prosecutors and judges with suspects based on genetic profiles matches. As Appel and colleagues (2018) argue, infrastructures are frequently the materialisation of hopes and promises. 'Whether they are being built or crumbling, infrastructures simultaneously index the achievements and limits, expectations and failures, of modernity' (Appel et al., 2018, p. 26). Because of this double index of achievements and failures, the infrastructure of forensic genetics in Brazil needed careful strategies to have its successful installation guaranteed. It could not rely solely on sociotechnical imaginary over science, technology, genetics and databases. One of these strategies became clear through the voice of the representative from the Gordon Thomas Honeywell (GTH) lobby firm.

Since at least 2008, pressure activities have been registered towards Brazilian experts and legislators from the GTH governmental affairs firm, lobby specialists hired by the biotechnology company Thermo Fischer to represent their interests in the expansion of forensic DNA databases around the world. In June 2008, before the agreement with the FBI, the president of the GTH gave two lectures in Brazil that focused on the strategy that, according to him, worked in the United States: 'pass the law and the money will come'. Brazil thus needed to prepare its DNA database law quickly if it wanted to benefit from the promise that the DNA database carried.

The presence of GTH lawyers and lobbyists could be observed in Brazilian forensic genetics seminars until at least 2014. One of the authors of this chapter had the opportunity to meet one of these lawyers and could observe and interact with him during these seminars, where several forensic geneticists approached him to share news about how the sampling inside prisons in each state was being prepared. In 2014, forensic experts were facing difficulties in getting access to prisoners' DNA samples. The lawyer was already well known among the Brazilian experts, frequently addressed by his nickname in a friendly tone, indicating that they had worked together and met several times. He expressed to forensic experts and one of the authors of this chapter his concern about the time it was taking to start sampling the Brazilian carceral population. Two years had passed since the DNA database law's approval, and the databases were still empty, even with the combination of a restricted mandatory inclusion criteria for *crimes hediondos* [serious offenses] and a broad category of 'violent nature against a person'[8] that should help justify and accelerate the process and give judges discretions to ask for more mandatory sampling of convicted individuals. The lobbyist approached representatives from each state to get information on how close each state was to being granted access to prisons to conduct DNA sampling from those eligible. The company he represented had sold the technology to equip the new laboratories set in place to feed the database network, thus generating millions of dollars in revenue. The success of the DNA database and the company he represented depended on having access to the third largest carceral population in the world.

In Brazil, the money did flow. Today, the country has 20 laboratories connected to the national network and the third largest, and growing, carceral population in the world (915,000 persons, 0.43% of the population).[9] This large carceral population and its data are expected to translate into a large revenue to companies following the projected expansions to the criteria of mandatory sampling, expansion that is always presupposed by the very legal frames that these companies present and enforce in Brazil. The everyday practical conditions to do so, however, have proved to be way more complicated than forensic experts and legislators expected and supposed.

A *peritos'* law: Situating the challenges to Brazilian DNA database law

When the DNA database technology became available in Brazil, the legal framework that would allow its use in criminal investigations was still to be established. This opened the usual regulation questions: Who should be in these databases? How should they be filled? How long should profiles be stored? Should the convicted be required to give samples or should they consent?

In 2011, the DNA database bill was presented to the Senate, and after a fast legislative process by Brazilian standards, it was passed into Law 12.654 in May 2012, authorizing mandatory DNA sampling and the storage of genetic profiles. The following year, the *Banco Nacional de Perfis Genéticos* (BNPG) [National Genetic Profile Database] was created. Brazil's forensic DNA database law prescribed that all individuals convicted of '*crimes hediondos*'[10] and crimes of a 'violent nature against a person' must submit to mandatory DNA sample collection. The genetic profile is collected and stored by state-level laboratories that feed the BNPG. The national database, then, compiles all profiles uploaded by the network of state laboratories through periodic automated uploads. These genetic profiles are retained for the crime prescription period,[11] and access to the national database information is restricted to authorised forensic experts who will communicate matches obtained by the automatic comparison of profiles (automated speculative search) to the state laboratories through a report.[12] After states' laboratories receive a report saying there's a match with another state database, the two state laboratories have to communicate with each other to see what kind of relation the profiles have. Police officers can have access to the database during investigations with an authorisation provided by a judge, but the automatic comparison provided by the CODIS software is the main form of obtaining a match.

Far from closing the discussion on how Brazil should use forensic DNA databases, the law intensified debates. Several criminal law scholars and some judges stated that the law was approved without broad legal experts and scholars consulting. As Dr Taysa Schiocchet, a legal scholar who coordinated a research report to the Ministry of Justice on forensic DNA database law (Schiocchet, 2012) said in an interview in 2014, the scarcity of public debate during legislative processes is not rare in Brazil. However, the restricted debate around forensic DNA databases in Brazil was an effect of the *peritos*' privileged position during the DNA database's law elaboration process.[13]

> Few people knew about the existence of the law. Few *juristas* [legal scholars] participated. Very few *juristas* participated in this process! The law didn't have a democratic discussion with society. For sure the whole society couldn't participate. How could we have done? Expanding the debate, bringing other experts, listening to other point of views. Our DNA law have only five articles! Measly five articles! Maybe this whole process could have been better
>
> *(Schiocchet, interview conducted in December 2014)*

Dr Schiocchet's account is not far from the perception of the *peritos* themselves. Gustavo, a forensic scientist for the federal police with more than fifteen years of experience in forensic genetics interviewed in Porto Alegre, said that other actors from the criminal justice system, such as *delegados* [police chiefs], prosecutors, lawyers, public defenders and judges had minimal participation during the introduction of forensic DNA databases in Brazil.

> All the initiative around the DNA database was taken exclusively by *peritos*. From building laboratories to structuring the database. The law itself was an initiative taken by the *peritos*. We went looking for legislators. We went looking for a Senator and said to him: 'look, we are going to present a law like this and this'. . . . And he liked the idea. We approached some congressmen and took them to get to know the CODIS in the United States. It was an initiative taken by the *peritos*. The participation of judges and prosecutors in the shaping of the DNA law was null. It didn't exist. When they got in? They got in [the debates] now, when they started to participate of the database administration committee.

Restricted debate on forensic DNA database law were pointed out by critics as one reason for the approval of 'a law full of unconstitutionalities' (Nicolitt, 2013). The main aspect that drew criticism was the mandatory collection of genetic samples from those convicted of a *crime hediondo*, violating the right of citizens to not self-incriminate.[14] This account sustained an appeal challenging the constitutionality of the DNA database's law submitted to Brazil's Supreme Court in 2016. The case began in April 2014, when a Minas Gerais's state prosecutor asked a judge to authorise the DNA sampling of a convicted felon who was granted probation. He was a suspect in another crime under investigation. The judge denied the request on the grounds that demanding a DNA sample from someone against their will was self-incrimination. The prosecutor appealed, and the decision was reversed. The state of Minas Gerais's Defensoria Pública [public defense office] then succeeded in taking the case to the Supreme Court. In May 2016, the Court accepted the case as having constitutional relevance, and in May 2017, the Supreme Court held a public hearing on the matter.

The hearing took place simultaneously with the Interforensics Seminar, the largest forensic sciences event in Brazil. Because of this, the hearing had the participation of prominent international and Brazilian forensic experts, such as the directors of the forensic DNA databases from United States and Germany. Alongside them, other Brazilian forensic experts, legal scholars and Debbie Smith, an activist from a support group for victims of sexual violence, addressed the Court as *amicus curiae*.[15] Douglas Hares from the FBI told the hearing judge that in his country, 225 cases were filed challenging the constitutionality of DNA laws in 42 states. These cases challenged regulations towards searches and seizures but also self-incrimination through mandatory DNA sampling. In all these cases, US courts have ruled that there was no valid legal reasoning against searches and seizures or self-incrimination. The director of Germany's DNA database addressed the size and success of British and German databases that, in Germany, have led to a match in every three new profiles included. The lawyer representing the Brazilian Academy of Forensic Sciences reviewed the situations that the Brazilian Supreme Court considers a violation of the right to not self-incriminate: 1) when the defendant is made to collaborate in an 'active way' to produce evidence and 2) through invasive (bodily) extraction of evidence. For him, mandatory DNA sampling doesn't qualify in these situations because it's a passive and non-invasive procedure through a mouth swab, allowing for forced mandatory collection. But if the Court still considered this a violation, the lawyer proposed the possibility of sampling DNA from the suspect's belongings without consent, something that was already authorised by the Supreme Court for other type of evidence. This demonstrates the importance of having the DNA database filled. For most Brazilian forensic experts, the DNA database's value depends on the quantity of profiles it contains. Starting the sampling in prisons with their large populations thus seemed to provide a convenient point of departure. Yet what stood and stands in the way of this convenient point of departure in Brazil is a combination of a troubled history dating back to the 1960s when Brazil was a military dictatorship and everyday contemporary prison practices and routines that escape state security forces' control.

Sampling inside prisons and dictatorship's torture memories

The mandatory DNA sampling, the main underlining subject of the hearing, was addressed again through questions by the state of Minas Gerais's public defense office attorney to the director of the BNPG. The director's presentation focused mainly on the low numbers the national database had accumulated over five years.[16] The director estimated that 70,000 persons in Brazilian prisons were convicted of *crimes hediondos*, and the validation of the constitutionality by the Supreme Court would help enforce DNA collection inside prisons. The public defense office's lawyer asked the director if she had participated in a DNA sampling inside a prison, how the collections were conducted and if the reason for the collection was explained to inmates. The director answered that she hadn't participated in one, but she could speak based on the experiences of her colleagues. She said that a sampling organised in the Catanduvas prison, Brazil's model federal maximum-security penitentiary, was arranged through the *Vara de Execuções Penais* [sentencing oversight court]. The director said that inmates were previously selected by judges and prison administrators in such a way that the forensic team arrived at the penitentiary ready to sample without knowing what had been said to the inmates. In an attempt to answer the public defense concerns, the director once again showed the pictures of a sampling conducted inside a prison in the north region of Brazil that she had used in her presentation. The forensic geneticist, properly dressed, stood in front of an inmate seated in a chair with his open mouth turned up towards her. Nothing in that picture suggested difficulties of sampling inside a prison. However, the concerns of the public defense lawyer remained: What would happen if the inmate refused to open his mouth?

This question lingered and was addressed one more time by the public defense after a talk given by a forensic physician representing the *Associação Brasileira de Medicina-Legal e Perícias Médicas* [Brazilian Legal Medicine Association], a group of forensic experts who sometimes dispute the authority of sampling genetic profiles. Addressing the medical doctor, the defense lawyer said that mandatory sampling inside prisons could result in the use of force or restraining tactics. The lawyer asked how the forensic physician expected these practices would be conducted inside prisons. The physician answered by citing the intriguing dilemma that his peers faced when discussing this matter:

> As legal physicians, we have a profoundly serious problem. The Resolution 1931 of the Federal Medicine Council forbids us to do that [collect DNA without consent]. It doesn't mean that the law forbids me to do that, but the Council does. We are, before anything, medical doctors. I can't do this kind of procedure inside a prison.
>
> *(Dr Jozefran Freire, 26 May 2017)*

For the physician, one solution to this dilemma could be the transfer of all persons to be genetically profiled to an appropriate examination facility. However, he recognised that the safety and public security risks involved could make it impossible. He went further and considered that even if this could be accomplished, what would happen if someone refused to open his mouth, even with a judge's order? 'What am I to do to make someone open his mouth? I won't be able to do this. I can't force him to do anything. We need the consent'. The public defense lawyer interrupted: 'He [the prisoner] will say no'. 'We won't do anything, then', responded the physician.[17]

The RIBPG's guidelines do include a document to inform judges on inmates' refusal to give a DNA sample, and the Brazilian forensic experts reject recourse to violence as a means to obtain a DNA sample. They agree with those concerned that if the sampling of an inmate were to be conducted by penitentiary agents, it would drastically reduce the credibility of the evidence. These samples would not carry the authority of forensic expertise and its association with science, technology and their protocols. They could carry, instead, association with the violence and deceit frequently used by Brazilian penitentiary agents to manage prisoners' everyday life. These practices would give defense lawyers more chances to raise doubts about DNA tests results, sampling credibility and chain of custody inside prisons since Brazilian police practices of tampering with evidence and crime scenes are well and largely known. Besides the effects on the credibility of the evidence produced from such sampling practice, these associations between genetic sampling and police practices inside prisons evoke references to the military dictatorship, risking the erosion of the credibility of forensic DNA databases as a whole.

Although the association between violence, deceit and prison practices is more frequently associated with police, during the Brazilian dictatorship, forensic expertise and evidence were used to cover up homicides conducted by the military regime. This history and these memories are not insignificant since forensic expertise had an important role, for example, in the case of the torture and murder of the journalist Vladimir Herzog on 25 October 1975. Forensic evidence and reports were forged by the military regime to make it look like a suicide. After Clarice Herzog's efforts to prove the murder of her husband, the complot between the torture unit (DOI-CODI), forensic experts and state prosecutor was exposed. Six days after Vladimir Herzog's murder, an ecumenic act was held in São Paulo with leaders from the Catholic, Jewish and Evangelic faiths, gathering thousands of people even under military surveillance and siege. It was an iconic moment towards the long process of rebuilding democracy in Brazil. Alongside the dictatorship history, cases such as the Carandiru massacre, where 111 surrendered inmates were executed by military police during a rebellion at Carandiru State Prison in 1991, many *chacinas* [massacres] and corruption cases involving police officers and high numbers of deaths caused by police action make the link between prisons, police forces and bodily samples with possible implications for criminal prosecution very sensitive. Brazilian forensic geneticists, therefore, need to navigate and relate to these memories very carefully in order to keep sampling practices inside prisons as far away from police forces as they can to better preserve the credibility of such a crucial aspect of forensic DNA databases.

The dilemma pointed out by the legal physician, therefore, remains even if the Supreme Court rules that mandatory DNA sampling is constitutional. As the public defense attorney expressed in the hearing, attentive and informed defense lawyers won't stop counseling their clients to refuse sampling, and the questions about how far to go to constrain someone and obtain a DNA profile are still far from resolved.

Expanding the mandatory sampling in Brazil: Consolidation of a DNA database model

As of September 2021, the Supreme Court hadn't manifested a decision. However, the now-former minister of justice Sergio Moro made proposals to expand the criteria for mandatory sampling in his Anti-Crime Law. This bill caused heated discussions in 2019 because it included an 'illicit exclusion' of police officers who killed while on duty if it was proven that the action was committed under 'fear, surprise or violent emotion'.[18] Concerning DNA databases, the bill proposed to expand mandatory sampling to individuals convicted for a *crime doloso*, a crime committed with intent. The bill also established that convicted individuals already incarcerated would be subject to mandatory sampling, something that wasn't inscribed in any law, leaving the responsibility for organising the sampling to forensic geneticists. The last aspect concerning the database was that it made the refusal to give a DNA sample a '*falta grave*' [severe fault] that would have the effect of cancelling sentence progressions and other benefits acquired by inmates during the execution of their sentences.

The Anti-Crime Law was proposed by a notorious judge responsible for a case that unsettled the country from 2014 onwards ('car wash').[19] As an icon of populist punitiveness, in which a new law promises to end crime through increasingly severe punishment, this bill was proposed in a context characterised by increasing authoritarian inclinations. As Lima and colleagues (2020) have argued, the expansion of authoritarian sensibilities among the Brazilian population is intimately related to the country's shortcomings in dealing with violent crime and the generalised fear of violence. They observed that authoritarian sensibilities and inclinations are more frequent in lower income groups, and the authors relate this to the precarity of social programs dependency, constant risk of losing social benefits and higher susceptibility to violence. In such a context, where harsh criminal laws are frequently viewed as the ideal way to deal with crime and violence, the Anti-Crime Law was received with support from different segments of Brazilian society, and the public debates centered on the polemic 'illicit exclusion'. Doing so, the widening of DNA database inclusion criteria towards less-offensive crimes went under the radar of public debate once again.

Brazilian forensic experts, legal experts and scholars expected an expansion of mandatory DNA sampling criteria, more or less in line with the more common trajectory of forensic genetic databases described by Robin Williams and Paul Johnson (2008). In such cases, initially, a database would typically be based on narrow inclusion criteria, usually homicides and sexual assaults, and later, while promising efficiency,[20] the database criteria would be expanded to more common and frequent crimes, such as property crimes and smaller infractions.[21] In Brazil, such expansion translated into the sampling of way more individuals than the 70,000 convicted for *crimes hediondos* that the director of the BNPG had estimated during the 2017 Supreme Court public hearing. In fact, this estimation has already proven very cautious, since the latest BNPG report (through May 2021) shows a total of 110,579 genetic profiles stored with 75.46% (83,439 profiles) being from convicted individuals and only 16.42% (18,152 profiles) from crime scenes.[22]

Thus, the expansion of the criteria for mandatory sampling implied the adoption of a DNA database model focused on convicted individuals. As Santos et al. (2013) demonstrated, the focus on crime scene samples can make smaller databases perform just as well as those with a large percentage of the population and fewer crime scenes profiles stored. The option for a convicted individuals' model, therefore, cannot be sustained based on an efficiency argument. In order to better understand Brazil's choice of this model, we have to go beyond the supposed effectiveness of the database and also beyond the interest of the biotechnology industry. If we include forensic experts' points of view on police organisation and investigation practices, we learn that the police infrastructure puts important constraints on the circulation of genetic profiles.

Unsettling infrastructures: Two police model and crime scene practices

During the period prior the approval of the forensic DNA database law, police work at crime scenes was a constant concern. It still is. Yet this concern is usually pushed aside by most forensic experts as just a matter of better training for police forces. What is ignored, however, is that even with police training to handle crime scenes according to forensic genetics standards, police infrastructures such as the two police model and the concentration of forensic services in state capitols will remain in place. Changing this mode of working is a much harder reform that both Brazilian forensic experts and biotechnology companies know they cannot address or demand. These infrastructure characteristics present significant threats to the project, filling the databases with crime scene profiles, making it an important factor in the process of establishing the database model and its focus on convicted individuals inside prisons. Relying solely on known individuals' genetic profiles, nevertheless, covers only part of the promise that the DNA databases carry.

The Brazilian forensic experts we interviewed in our research frequently insisted on highlighting that most violent crimes are committed in the streets of the city's poorest neighborhoods. These are typically crime scenes where the police arrive after the victims' family or curious people, making it harder to secure evidence in an optimal manner. Crime scenes are wittingly or unwittingly tempered with. Besides the challenges of large urban centers, forensic experts also point to the distances they have to travel to get to countryside crime scenes since the vast majority of forensic services, especially the forensic genetics services, are located in state capitals, and police officers are not authorised to work a crime scene, beyond its isolation, until it is 'cleared' by forensic experts.[23] For Walter, a forensic expert at Rio Grande do Sul's *Instituto-Geral de Perícias* (IGP) [Forensic Institute], in these situations, crime scene preservation is even more difficult.

> This culture of crime scene preservation is something that the IGP has been promoting for the last 12 years. It established the following rule: we only go to crime scenes that were isolated and guarded. If you arrive at a scene and it hasn't been isolated, it's a scene compromised. You don't have much to do there. Of course, in exceptional cases you end up answering the call, because of a political pressure or something like that. But in general, we don't answer for crime scenes that haven't been isolated and guarded.
>
> *(Interview conducted in December 2014)*

As has already been established by science and technology studies accounts (Lynch et al., 2008; M'charek, 2008), isolation and preservation are not only a matter of maintaining crime scene conditions for a larger collection of samples and evidence. It is the first in a series of procedures that take part in the

DNA evidence's economy of credibility and in the evidence enactment itself. As M'charek argues,

> DNA is more than just biological material. The DNA is inextricably bound up with all those procedures and techniques necessary to be able to use it as means of identification. Without those procedures and techniques, you do not have DNA to start with (maybe a T-shirt with blood on it, but no more than that).
>
> *(M'charek, 2016, p. 40)*

This focus on the procedures and techniques involved in forensic DNA evidence led M'charek to conclude that the movement of a biological sample found at a crime scene through the mediation of police investigators, forensic geneticists, laboratories, PCR, genetic sequencers, forensic reports and judicial proceedings together make DNA evidence. In these movements, identities and contexts are made.

> The route of the biological material from crime scene to lab and out again comprises therefore more than the transmission of material and information. Along that route a biological trace is made into DNA evidence, a police officer becomes a forensic sleuth, and a genetic researcher becomes an expert witness. The various actors together make DNA what it is: forensic evidence. But also the other way around, the DNA that circulates between them makes them what they are; all are more than the title of their function would suggest.
>
> *(M'charek, 2016, p. 41)*

Circulations are, therefore, performative, and Brazilian forensic experts are aware of and concerned about the challenges presented to the circulation of crime scene samples that should eventually lead to credible forensic evidence. The particularities of a large country like Brazil, with its main state's institutions concentrated in urban areas, are not the only concerns involving crime scene preservation. The model of two police forces, one military and another civil, adds significant complexities to the practices required to make crime scenes' genetic profiles circulate with credibility through the infrastructure of forensic genetics.

In Brazil's police organisation, the military police are the first public security actors to arrive at a crime scene and to act upon it, but without any prerogatives to conduct criminal investigations. The *peritos* expect that military police actions will be directed towards crime scene isolation, bearing in mind the subsequent procedures needed to make biological material into DNA evidence. The civil police, on the other hand, are responsible for investigations and judicial work. However, as a bureaucratic institution, it frequently responds too slowly to criminal acts, leading to lower-quality information and greater difficulties in investigations (Soares, 2007). This dual system has been seen as a main obstacle in Brazil's efforts to reduce violent crime and improve crime resolution rates.[24]

The territorial distribution of forensic institutions and police institutional organisation's impact on crime scene preservation were expected to present daring challenges to the introduction of forensic DNA databases in Brazil. If these characteristics already presented problems to circulating other forensic evidence, when DNA arrived on the Brazilian technolegal scene, it became even more daring. This technology was introduced through the mediation of the FBI, including the CODIS training that Brazilian *peritos* received from the FBI and adoption of certain practices, such as checklists for the profiles to be uploaded to the database, which is a crucial aspect of establishing the credibility of the chain of custody and the DNA evidence. This kind of checklist, however, becomes hard to comply with when crime scenes frequently slip away from *peritos*' control. Training programs for both police forces are offered to reduce these problems around crime scene preservation. The organisation of policing, however, is not expected to change anytime soon, perpetuating some of the expected challenges that surround the introduction and stabilisation of forensic DNA databases in Brazil.

Concluding thoughts: 'Scripts' of the Brazilian DNA database model

The Brazilian forensic DNA database's model of sampling known individuals emerges as a combination of international collaboration with the FBI, biotechnology companies' interest in sampling a large prison population, promises of crime deterrence from the use of genetic science and technology and the particularities of Brazilian forensic and police infrastructure, presenting known challenges to forensic experts in dealing with DNA evidence at crime scenes. This model, however, began to face unexpected difficulties once it was time to fill the database with individual samples collected inside prisons, as the head of the BNPG made clear at the Supreme Court public hearing. While the challenges to be faced around crime scenes and valid DNA evidence were anticipated by Brazilian forensic, legal and police experts, the mandatory DNA sampling of known and convicted subjects was very quickly reframed based on the assumption that prisoners were easily available to be sampled. According to the experts we spoke to, embracing a model for the DNA database that is based on known individuals – convicted prisoners – was deemed easier and more viable than one that focused on crime scene samples. Moreover, the model of sampling known convicted individuals more than crime scenes has always been considered the best, fastest and cheapest way to insert Brazil's forensic science infrastructure in the 'modern' and 'scientific' practices of policing through DNA databases. This assumption indicates an approximation of the larger process of making public bodies out of individual bodies through the mediations of law, technology and science described by Victor Toom (2012). Through the notion of the forensic genetic body, Toom analysed how the mastery over individuals' bodies by state authorities and the loss of civil rights and entitlements by these individual bodies become conditions of forensic genetics' knowledge, making bodies and bodily substances increasingly the center of contemporary policing practices and technologies.

Following this trend and these assumptions, it remains to be seen how two important aspects develop in the Brazilian technolegal setting: 1) how the sampling inside prisons will be stabilised and generalised beyond the few and small-scale samplings following initiatives and agreements in each state and 2) whether the sampling strategies inside prisons will include the participation of police and prison employees, risking jeopardising the samples' credibility, or whether they will be conducted by forensic experts, with full knowledge of the procedures provided to inmates and their defense lawyers.

The forensic DNA database technology cannot be considered isolated from the mediations and associations that contributed to its arrival and stabilization in Brazil's legal and infrastructural setting in a particular way. As Antina von Schnitzler (2013, 2015) argued, when technologies depart from their production context to travel to another, they become even more unstable, opening spaces for disputes and negotiations on new uses, regulations and meanings that are part of the process of their stabilization. However, as Madeleine Akrich (1992) reminds us, technologies also carry 'scripts' on the distribution of agencies, relations and subjects they interact with. These scripts are the work of the technology designers to inscribe a particular vision or conception of the world that 'define[s] actors with specific tastes, competences, motives, aspirations, political prejudices, and the rest, and they assume that morality, technology, science, and economy will evolve in particular ways' (Akrich, 1992, p. 208). It means that technologies and infrastructures set frameworks that embody politics and moralities in their very design.

These ideas are particularly relevant to understanding the difficulties that Brazilian forensic experts had in filling the database with genetic profiles sampled from incarcerated individuals. When the forensic DNA databases arrived in Brazil, accompanied by the United States and United Kingdom model of sampling known individuals (i.e., suspects of a crime), the stakeholders responsible for its introduction inscribed the only model supposedly available to Brazil based on the availability of prisoners' bodies. The bodies of incarcerated subjects were assumed to be at the disposal of forensic experts. Their bodies are contained and lives are managed by a state institution in an infrastructure that was deemed easily accessible to other state actors, such as forensic experts, through negotiations with judges and prison wardens. The Brazilian prison system, however, holds particularities and complexities way more difficult than one can expect from a distance and beyond what we could describe and address here. From the offices and laboratories, the prison can easily be framed as an infrastructure at a state actor's disposal. But getting closer to Brazilian prisons and knowing their intricate everyday practices can help us appreciate that forcing a prisoner to open his mouth to provide a DNA sample and to circulate it to police genetic laboratories can be much more delicate. Brazilian police, forensic and state practices towards prisoners cannot be detached from their brutal history, connected to the military dictatorship, the precarious conditions that can be observed today, the history of internal conflicts between police and criminal rival groups and the control that criminal organisations have of many prisons' everyday life. As Karina Biondi's ethnographic work (2010) on organised crime in Brazilian prisons has shown, criminal organisations such as *Primeiro Comando da Capital* (PCC) are deeply involved in the management of everyday life in Brazilian prisons and extend their territory of control far beyond the prison walls, into the suburbs of large urban centers such as São Paulo. This doesn't mean that the inmates have total control over Brazilian prisons. But it means that, although the number of genetic profiles stored began to rise in 2019, prisoners still are a collective of subjects who can't be forgotten or assumed irrelevant. Thus, when Brazilian forensic geneticists want to circulate DNA through the forensic infrastructure in order to do individuals and suspects through the mediation of forensic DNA databases, they may encounter not one individual who does not want to open his mouth for swabbing but a collective shaped by the work of PCC and other organised groups. Negotiations with the inmates might be necessary to enable the collection and eventual circulation of their DNA, even if the mandatory sampling is considered constitutional by the Brazilian Supreme Court. Moreover, a prisoner's DNA could become an object of negotiation and traded for sentence revision, thus reducing his time in jail. This negotiation could involve a better understanding of the implications of DNA sampling for inmates since, as Richter and Fonseca (2018) describe, it's not clear that prisoners are being well informed about DNA sampling procedures and their consequences. Otherwise, violence and deceit may characterise the conduct of mandatory DNA collection inside Brazilian prisons, an alternative that will flood the courts with appeals that has the potential to jeopardise the project of establishing the largest forensic DNA database network outside the United States as a promise to halt the urgent problems around crime and police investigations in Brazil.

The possibility of turning prisoners' DNA into an object of negotiation could be viewed as a technolegal innovation in Brazil. It has the benefit of increasing the number of samples stored in databases without using deceit and misinformation practices, and recognising prisoners' rights and autonomy over their genetic information. At the same time, it gives inmates better information on the significance and consequences of having their genetic profiles stored and identifiable by routine database comparisons. How this access to knowledge about forensic DNA and the consequences of the DNA databank for prisoners will affect the growth of the database, and thus how it will further shape technolegal worlds in Brazil, remains an open question.

Notes

1 The authors would like to thank the editors for the inspiring comments, suggestions, critiques and patience throughout the process of conceiving and delivering this chapter during COVID-19 pandemic times.

2 *Diário Oficial da União* section 3, number 110 from 12 June 2009, page 81. The notification of the agreement does not include any details on the terms of the agreement. Through the law of government information access, we were able to obtain details on the agreement. The FBI commits to the training and support of Brazilian forensic experts and updates to CODIS system, and Brazilian police forces commit to accepting updates to the software that are indicated by the FBI. Both parties agree that they won't

have access to each other's databases, except through the regulations that are directed to this specific matter by each institution.

3 The 2012 report can be accessed here: www.justica.gov.br/sua-seguranca/seguranca-publica/analise-e-pesquisa/download/estudos_diversos/2diagnostico-pericia-criminal.pdf.
4 For an example in the northeast region, see Richter and Fonseca (2018).
5 Defined as 'ways in which scientific and technological visions enter into the assemblages of materiality, meaning, and morality that constitute robust forms of life' (Jasanof, 2015, p. 4), this notion helps us highlight how the promise of a new infrastructure of forensic genetics in Brazil involved the allure of science and technology and their acritical acceptance (Fonseca, 2013; Machado, 2004) as an effect of this particular sociotechnical imaginary.
6 The report can be found here: https://www.ipea.gov.br/atlasviolencia/publicacoes/47/atlas-da-violencia-2017. Last accessed 28 September 2022.
7 According to the FBI, the United States registered 16,214 homicides in 2019, and according to the British Office for National Statistics, the United Kingdom registered 617.
8 The fact that this crime typification doesn't exist in Brazilian penal law led to serious rumors that the legislation itself was written by the lobbying firm mentioned earlier and given to legislators to be submitted to the Senate. This is not an uncommon practice in Brazil as revealed with strong evidences by The Intercept Brasil in a report in which they show that 292 out of 850 (34.3%) amendments to a Constitutional bill on labor reform were submitted directly from lobby and bank computers. See https://theintercept.com/2017/04/26/lobistas-de-bancos-industrias-e-transportes-quem-esta-por-tras-das-emendas-da-reforma-trabalhista/. Last accessed 27 January 2020.
9 According to the Prison Monitoring National Database, maintained by the National Justice Counsel, 915,620 persons have been deprived of liberty in Brazil. Among them, 412,742 (45,07%) are under arrest provisionally and have not been convicted yet. https://portalbnmp.cnj.jus.br/#/estatisticas. Last accessed 18 November 2021.
10 Comparable to a felony in the United States.
11 The longest time for a crime prescription in Brazil is 20 years.
12 In state-level laboratories, only one forensic geneticist is responsible for accessing and updating the CODIS system.
13 For a discussion on restricted democratic debates on forensic DNA databases' legislation and organization, see Toom et al. (2019); Prainsack and Toom (2013), McCartney et al. (2011) Machado and Granja (2018).
14 Critiques were also directed towards violations of bodily information autonomy, privacy and informed consent. See especially Schiocchet (2015) and Schiocchet and Garrido (2018) for a compilation of these critiques.
15 Both authors were involved in a research group that participated as *amicus curie* in this public hearing and participated by speaking on it and through reports that were included in the proceedings of the case.
16 In May 2017, the National Genetic Profiles Database (BNPG) had 8,916 genetic profiles. 4,971 were crime scene samples, and 1,687 were samples from convicted inmates. See www.justica.gov.br/sua-seguranca/seguranca-publica/ribpg/relatorio. Last accessed 20 September 2020.
17 For a similar discussion but with different outcome, see Toom (2010).
18 This proposal was later taken off the bill.
19 The 'car-wash' case began in March 2014 as investigations into money laundering in gas stations involving small-scale politicians. The investigations took on more relevance when they expanded into corruption cases in the state oil company Petrobras and large private construction companies. These investigations were used to promote the responsible judge to a 'national hero' who would end corruption. The operation and its main characters became engaged in the destabilisation of left-leaning governments and politicians, with heavy participation in former president Dilma Rousseff's impeachment. The case took an even more direct role in Brazilian politics when the investigations focused almost exclusively on former president Luiz Inácio Lula da Silva. His convictions later

prevented him from running against Jair Bolsonaro in the 2018 presidential election. As a form of 'reward' for the work done by the judge in the case and in the elections, Bolsonaro nominated the judge to the Ministry of Justice in 2019. Today, all prosecutions and indictments made in these investigations have been nullified by the Supreme Court, given the leaks that emerged exposing communications between the judge and the Federal Prosecution Office, showing how they coordinated the investigation work, sentences, political strategies and motivations.

20 For important and insightful reviews of the debates over the effectiveness of DNA databases, see Amankwaa and McCartney (2019); Toom (2012).

21 See also Duster (2004).

22 The latest report can be found at www.justica.gov.br/sua-seguranca/seguranca-publica/ribpg/relatorio/xiv-relatorio-da-rede-integrada-de-bancos-de-perfis-geneticos-ribpg.pdf/view. Last accessed 26 September 2021.

23 Police officers do collect artifacts from crime scenes and send them to forensic laboratories to be analysed. But the evidence that emerges from this does not carry the same forensic authority and credibility, especially when it involves forensic genetics expertise and technologies, since it did not result from a full forensic crime scene examination.

24 In 2011, the Ministry of Justice and the National Justice Council led a task force to identify all unresolved homicide cases between 2007 and 2011 and to try to indict suspects when possible. The task force found 134,944 active and unsolved homicide cases around the country and was able to close 43,123 (31.95%) of the total. The task force only indicted a suspect in 8,287 cases (6.1% of the total). The report is available at www.cnmp.mp.br/portal/images/stories/Enasp/relatorio_enasp_FINAL.pdf.

References

Akrich, M. The de-scription of technical objects. In: Bijker, W.; Law, J. (Eds.). *Shaping Technology/Building Society. Studies in Sociotechnical Change*. Cambridge, MA: MIT Press, 1992, p. 205–224.

Amankwaa, A. O.; McCartney, C. The effectiveness of the UK national DNA database. *Forensic Science International: Synergy*, n. 1, 2019, p. 45–55.

Appel, H.; Anand, N.; Gupta, A. Introduction: Temporalities, politics, and the promise of infrastructure. In: Anand, N.; Gupta, A.; Appel, H. (Eds.). *The Promise of Infrastructure*. Durham and London: Duke University Press, 2018, p. 1–38.

Biondi, Karina. *Junto e Misturado. Uma etnografia do PCC*. São Paulo: Terceiro Nome, 2010.

Duster, Troy. Selective arrests, an ever-expanding DNA forensic database, and the specter of an early twenty-first-century equivalent of phrenology In: Lazer, D. (Org.). *The Technology of Justice: DNA and the Criminal Justice System*. Cambridge, MA: MIT Press, 2004, p. 315–334.

Fonseca, C. L. W. Mediações, tipos e figurações: reflexões em torno do uso da tecnologia de DNA para identificação criminal. *Anuário Antropológico* [Online] 2013, p. 9–33.

Jasanoff, S. Future imperfect: Science, technology, and the imaginations of modernity. In: Jasanoff, S; Kim, S.-H. (Eds.). *Dreamscapes of Modernity. Sociotechnical Imaginaries and the Fabrication of Power*. Chicago: University of Chicago Press, 2015, p. 1–33.

Lima, R.; Jannuzzi, P.; Moura Junior, J.; Almeida Segundo, D. Medo da violência e adesão ao autoritarismo no Brasil: proposta metodológica e resultados em 2017. *Opinião Pública*, Campinas, v. 26, n. 1, 2020, p. 34–65.

Lynch, M.; Cole, S.; Mcnally, R.; Jordan, K. *Truth Machine. The Contentious History of DNA Fingerprint*. Chicago: University of Chicago Press, 2008.

Machado, H. Dilemas e paradoxos da cientifização da justiça versus a judicialização da ciência—o caso do ADN. *Revista Manifesto*, n.°7, 2004, p. 64–75.

Machado, H.; Granja, R. Ethics in transnational forensic DNA data exchange in the EU: Constructing boundaries and managing controversies. *Science as Culture*, v. 27, n. 2, 2018, p. 242–264.

McCartney, C.; Wilson, T.; Williams, R. Transnational exchange of forensic DNA. Viability, legitimacy, and acceptability. *European Journal on Criminal Policy and Research*, v. 17, n. 4, 2011, p. 305–322.

M'charek, A. Silent witness, articulate collective: DNA evidence and the inference of visible traits. *Bioethics*, v. 22, n. 9, 2008, p. 519–528.

M'charek, A. Circulations: A new object for an anthropology of science. *Inaugural Lecture*, n. 564, 2016, p. 35–61.

Misse, Michel. *Avaliação da Formação e da Capacitação Profisional dos Peritos Criminais no Brasil.* Brasília: SENASP, Abril 2006.

Nicolitt, André. Banco de dados de perfis genéticos (DNA): As inconstitucionalidades da Lei 12.654/2012. *Boletim IBCCRIM*, v. 245, 2013, p. 15.

Prainsack, B.; Toom, V. Performing the union: The Prüm decision and the European dream. *Studies in History and Philosophy of Biological and Biomedical Sciences*, n. 44, 2013, p. 71–79.

Richter, V.; Fonseca, C. Desafios éticos da genética forense no Brasil: sentidos e práticas em debate. In: Schiocchet, T.; Garrido, R. G. (Org.). *Bancos de perfis genéticos para fins de persecução criminal.* 1ed. Rio de Janeiro: Multifoco, 2018, v. 2, p. 23–50.

Santos, F.; Machado, H.; Silva, S. Forensic DNA databases in European countries: Is size linked to performance? *Life Sciences, Society and Policy*, v. 9, n. 1, December 2013, p. 1–13.

Schiocchet, T. *Banco de perfis genéticos para fins de persecução criminal.* Série Pensando o Direito, vol. 43. Brasília: Ministério da Justiça, 2012.

Schiocchet, T. (Ed.). *Bancos de perfis genéticos para fins de persecução criminal: análise interdisciplinar e em direito comparado.* Rio de Janeiro: Multifoco, 2015.

Schiocchet, T.; Garrido, R. G. (Eds.). *Bancos de perfis genéticos para fins de persecução criminal: práticas periciais e impactos jurídico-sociais.* Rio de Janeiro: Multifoco, 2018.

Soares, L. E. *Meu casaco de general. 500 dias no front da segurança pública do Rio de Janeiro.* São Paulo: Companhia das Letras, 2000.

Soares, L. E. Política nacional de segurança pública: histórico, dilemas e perspectivas. *Estudos Avançados*, v. 21, n. 61, 2007, p. 77–97.

Toom, V. Inquisitorial forensic DNA profiling in the Netherlands and the expansion of the forensic genetic body. In: Hindmarsh, R.; Prainsack, B. (Eds.). *Genetic Suspects. Global Governance of Forensic DNA Profiling and Databasing.* Cambridge: Cambridge University Press, 2010, p. 175–196.

Toom, V. Bodies of science and law. Forensic DNA profiling, biological bodies, and biopower. *Journal of Law and Society*, v. 39, n. 1, 2012, p. 150–166.

Toom, V.; Granja, R.; Ludwig, A. The Prüm decisions as an aspirational regime: Reviewing a decade of cross-border exchange and comparison of forensic DNA data. *Forensic Science International: Genetics*, n. 41, 2019, p. 50–57.

Von Schnitzler, A. Traveling technologies: Infrastructure, ethical regimes, and the materiality of politics in South Africa. *Cultural Anthropology*, v. 28, n. 4, 2013, p. 670–693.

Von Schnitzler, A. *Democracy's Infrastructure. Techno-Politics and Protest After Apartheid.* Princeton: Princeton University Press, 2015.

Williams, R. DNA databases and the forensic imaginary. In: Hindmarsh, R.; Prainsack, B. (Eds.). *Genetic Suspects. Global Governance of Forensic DNA Profiling and Databasing.* Cambridge: Cambridge University Press, 2010, p. 131–152.

Williams, R.; Johnson, P. *Genetic Policing. The Use of DNA in Criminal Investigations.* Cullompton: Willan Publishing, 2008.

4

FORENSIC DNA ANALYSIS AND DATABASE GOVERNANCE IN GHANA

Aaron Opoku Amankwaa and Judith Amankwa Addo

Introduction

Case Study: The 'Takoradi Missing Girls'

On different days in 2018, three young women and a girl set out in the morning for the day's activities. In the late hours of the day, their families became worried about their whereabouts and reported their disappearance to the Ghana Police Service. The families of the missing women later received calls from unknown individuals demanding ransom for their release, confirming that they had been kidnapped. The police traced the calls, leading to the arrest of Samuel Udoetuk-Wills, a Nigerian national, in December 2018. However, days after his arrest, he escaped from police custody. In the course of the investigation, in August 2019, the police discovered the personal belongings of one of the girls and some skeletal remains in Udoetuk-Wills's hideout. Based on a DNA test by the forensic unit of the Ghana Police Service, it was confirmed that the remains belonged to the four missing persons. The police arrested another suspect, John Oji, a Nigerian national, and both suspects were convicted of the murders of the four victims and sentenced to death by hanging (Aklorbortu and Dzodzegbe, 2021). Following the investigation and conviction of the two suspects, the families of the missing victims requested copies of the DNA test results and demanded an independent DNA test (Agyei Annim, 2021).

DOI: 10.4324/9780429322358-6

There is limited information about forensic DNA profiling and databasing in Africa. According to the INTERPOL global DNA profiling survey results (2019), 11 out of 53 African countries responded 'yes' to the question of whether they deployed DNA profiling (INTERPOL, 2019). Seven of these countries had also set up a DNA database, and two reported that such a database was planned. Twelve other countries responded to the survey but did not authorise the public release of the information provided. This chapter narrates developments in Ghana, one of the 11 African countries that have commenced using DNA profiling in the criminal justice system, but not without critical challenges such as illustrated in the case study referenced earlier.

Ghana began using forensic DNA analysis in 2011 when the Ghana Police Service Forensic Science Laboratory (FSL) was refurbished under the European Union-Ghana Police Project (Amankwaa et al., 2019). Through this project, about €3 million were invested in the development of the FSL.[1] The FSL is made up of five main units, including a DNA department, and the Accra-based laboratory is the main provider of forensic services to all 17 police regions in Ghana. In addition to the services provided by the FSL, other state institutions/agencies and private organisations offer forensic services to the police. These include the Ghana Standards Authority, DNA Diagnostic Centre (DDC) Ghana and some government and private hospitals. The capacity of the FSL to support police investigations and prosecutions has been criticised by Amankwaa et al. (2019), highlighting the need for investment in the central laboratory and the establishment of regional forensic centres through public-private partnerships and collaboration with university and hospital laboratories to aid the police in crime investigation and prosecution.

The governance of the use of forensic DNA in Ghana can be described as an *unregulated technolegal world*. The current regime is characterised by the lack of a dedicated legislative framework, robust quality management systems and independent statutory bodies to oversee the collection, retention, use and destruction of DNA records for policing purposes. In this chapter, we review the governance of DNA analysis as an emerging forensic practice in Ghana and the key issues associated with the current regime. There is presently no published review of the operation, growth, governance and performance of the DNA analysis unit of the FSL. The first part of the chapter provides a concise overview of the legal basis of the EU-Ghana Police project, setting the context of developments in Ghana, the key conditions of the EU aid and the implementation of the agreement. In the second part, we examine existing research on forensic DNA analysis and its application in Ghana since 2011. The third part discusses the trajectory of the development of a forensic DNA database and its associated issues. We then assess the impact of the use of forensic DNA analysis in criminal investigations, focusing on two publicly available case studies. Further, the representations of forensic DNA analysis and databases in the Ghanaian media is discussed, highlighting the key debates on the collection, retention, destruction and use of DNA for policing purposes. The final part of the chapter discusses the key principles governing the use of forensic DNA in criminal investigations and how these principles are being applied in Ghana. We demonstrate that the emerging *technolegal world* in Ghana lacks a robust governance framework to protect citizens and produce public trust in the police and forensic science. We end with a reflection on the status of DNA profiling and databases in Ghana, identifying gaps in the technology, research, policy and legislation, with recommendations to improve forensic genetics practice in Ghana.

The EU-Ghana Police project

The legal basis of the EU-Ghana Police project was the Cotonou Agreement, which was signed between the European Community and the African, Caribbean and Pacific (ACP) Group in June 2000 (ACP and European Community, 2000). The aim of the agreement, which ended in December 2020, was to promote reduction of poverty, sustainable development and economic integration among partner countries. The Cotonou Agreement enshrined four key principles that defined the partnership between ACP countries, the European Union and EU Member States. These principles included equality of all partners and ownership of development strategies, independence of ACP countries to determine development policies, public-private cooperation in the implementation of development strategies and conditionality of development needs to be based on countries' level of development.

Out of many billions of euros in European aid to ACP countries, the EU budget support for Ghana was estimated at €224 million (DEUG, 2012). The Ghana Police Service FSL was funded by the Ninth European Development Fund (EDF), with a total of €3 million to support crime fighting and improve security in the country. The areas of support for the FSL included construction, the supply of equipment (EuropeAid/129238/D/SUP/EU), software and forensic resources, vehicles and training (MOFEP, 2010, 2009a, 2009b). Currently, the scope and extent of the utilisation of the FSL resources are opaque, with limited information about the performance and subsequent development of the laboratory in the public domain. These issues raise questions about the accountability and transparency of the operation of the FSL.

Research on forensic DNA analysis in Ghana

Developmental and validation studies of forensic methods and procedures are crucial to demonstrate their reliability and thus usefulness as evidence in legal practices. These studies are furthermore required by established international standards (ILAC, 2014). Research output on forensic DNA analysis in Ghana is low, a gap that reflects the inadequate investment and prioritisation of forensic science in the last decade (Amankwaa et al., 2019; Koomson et al., 2019). To inform this chapter, a search was conducted on three academic databases – PubMed, ScienceDirect and Web of Science – to identify research work related to forensic DNA analysis in Ghana since 2011. The academic database search was conducted in August 2020.[2] A further literature search was conducted on the Ghana Police Service website and the two main universities in Ghana that are currently offering degrees and professional training in forensic science: Kwame Nkrumah University of Science and Technology and University of Cape Coast (Amankwaa et al., 2019). Additionally, a research inquiry questionnaire was sent to the FSL to obtain information about the operation and performance of the DNA laboratory. The literature search identified 99 hits from the three academic databases: 11 from PubMed, 4 from Web of Science and 84 from ScienceDirect. Overall, 12 relevant primary research publications were identified following the screening of the database hits, examining the references of included papers and searches on the police and the educational institution websites. No validation studies of DNA profiling methods and procedures employed by the FSL were identified, including the validation of specific multiplex systems in casework.

Multiplex systems in forensic casework

Modern DNA typing is mainly based on autosomal STR analysis, which is strongly supported by continuous research in genetics, technological advancement and the development of quality standards (Butler, 2012). In many jurisdictions across the world, there are set standards or common practices on the number and specific STR markers for forensic casework. This practice ensures consistency and cross-comparison of DNA profiling results across forensic laboratories in the same jurisdiction. For example, in England and Wales, DNA-17 profiling is the current multiplex system used by forensic science providers in casework (FIND Strategy Board, 2019). The DNA-17 system comprises all the 16 common European markers and 15 markers from the CODIS loci, allowing international comparisons.

In Ghana, there is no 'nationally approved' multiplex system or standard set loci for forensic casework, which suggests that different laboratories may use multiple systems for forensic DNA investigations. A standardised set of loci (e.g., the European Standard Set [ESS] loci of twelve core markers) is required to ensure the harmonisation of databases and the exchange of forensic DNA data (Prainsack and Toom, 2013). The current 'open status' in Ghana, however, may lead to complications in the comparison of DNA results from crime scenes and individuals. For example, it may not be possible to detect full profile matches if different multiplex test kits with different sets of markers are used for the analysis of biological samples from crime scenes and individuals. Further, the absence of a common set of loci may hinder the development of a common forensic DNA database to support police investigations (ENFSI DNA Working Group, 2017).

The evidence from the literature suggests that the GlobalFiler multiplex system (Wang et al., 2015) and other DNA-24 systems have been used in research and casework by scientists affiliated with the FSL (Afrifah et al., 2020; Badu-Boateng et al., 2018; Kofi et al., 2020). The DNA-24/GlobalFiler includes all the DNA-17 loci and the CODIS loci. The additional loci are the DYS391 and Y indel, which are specific to the Y chromosome. The DNA-24/GlobalFiler system was adopted by Scotland for casework in 2015 (BBC News, 2015). The DNA-24 profiling system is beneficial because it provides an opportunity to compare DNA results both nationally and internationally. The system is also superior to current existing multiplex systems because it offers high discriminatory power and sensitivity in DNA analysis. The advantages of the DNA-24 systems suggest a need for a comprehensive evaluation and validation of the multiplex system for application in forensic casework in Ghana. Following such evaluations, the system may be adopted as a common standard for forensic STR typing.

In addition to adopting a common multiplex system, there is a need to develop appropriate STR allele frequency databases to support the estimation of the significance of matching profiles in the population. This is important because DNA profiling technology only targets very few specific segments or markers of the entire genome of an individual. Hence, DNA profiles are not 'unique', making the interpretation of matching profiles a probabilistic evaluation exercise (M'charek, 2000). Population frequency databases provide information about the frequency of each possible STR allele or genotype in a specific population. This ethnic-specific data allows forensic scientists to estimate the random chance of occurrence of a profile in the relevant subpopulation. Ghana is a multi-ethnic society, with more than ten different ethnic groups. The major ethnicities include the Akans (48%), Mole-Dagbon (17%), Ewe (14%), Ga-Dangbe (7%) and others (14%) (CIA, 2020). To calculate a more accurate random match probability of profiles generated from crime scene traces and subject samples, forensic DNA scientists must have access to a representative allele frequency database (M'charek, 2000). In our literature search, we found six publications related to the STR allele frequencies, SNP and mtDNA haplotype frequencies of ethnic groups from Ghana (Fendt et al., 2012; Kofi et al., 2020; Poetsch et al., 2011; Sanchez-Faddeev et al., 2013b, 2013a; Wepeba, Iyengar, and Goodwin, 2019). These studies provide information that can be used for calculations in forensic casework to determine the significance of DNA evidence.

There is currently an ongoing project at the University of Central Lancashire investigating the allele frequencies of the GlobalFiler STR loci among four major ethnic groups from Ghana: 282 Akans, 250 Ewes, 262 Ga-Dangbe and 253 Mole-Dagbon and Northern minority ethnic groups (Wepeba, Iyengar, and Goodwin, 2019). The results of this study will provide new STR population frequency data for the main ethnic groups. The availability and accessibility of this data will significantly enhance the quality of forensic DNA casework in Ghana as well as enhancing the transnational exchange of DNA data. It is not clear to what extent the FSL utilise the existing population frequency databases and how DNA match evidence is interpreted and reported to the court. This information is wanting in available court proceedings and police public reports/communications in cases involving DNA (*Asante v The Republic*, 2017; Mensah, 2019).

National forensic DNA database in Ghana

A national forensic DNA database is a national intelligence database containing DNA profiles from known individuals and profiles generated from crime scene traces. In stranger offences (i.e., where the perpetrator is unknown), crime scene profiles can be searched in the database to identify potential suspects or unknown individuals who may be linked to the investigation. Currently, Ghana has no national electronic database of forensic DNA data (INTERPOL, 2019). Although there is no dedicated legislation on the collection, retention and use of DNA for policing purposes, the Data Protection Act 2012 categorises DNA information as special personal data that can be used in the administration of justice where necessary (Parliament of the Republic of Ghana, 2012). The existing regime implies that there is no possibility of forcing a suspect to provide a bodily sample for DNA analysis or storing forensic DNA data without a justification of its relevance in the case. However, an initial trajectory towards the creation of a database at the FSL has been established with limited statutory oversight, an approach described as 'rogue' databasing, in which DNA is collected legally, but there is no statutory basis for its retention (Murphy, 2013). The 2016 Interpol Survey on DNA profiling suggests that some DNA records from casework are being retained by authorities (INTERPOL, 2016). It is not clear whether profiles are held in an electronic format or hard copies or whether adequate measures are in place to assure the quality, security, legality and integrity of DNA samples and profiles.

To date, there has been no adequate parliamentary or public debate or research into public views on the acceptability of forensic DNA and databases for policing purposes. This lacuna has been recognised by other authors and stakeholders and its mending as an essential requirement to ensure that sensitive policing technologies, such as DNA databases, are acceptable to citizens (Science and Technology Committee, 2005; Wallace, 2006). The implications of the gaps outlined here include possible legal challenges on the privacy of individuals and the efficiency of the police in detecting or solving crime. Table 4.1 provides a breakdown of the DNA profiles held in Ghana as of 2016 based on the available Interpol report. The report provides no clarification on the custodianship of the DNA profiles or where the profiles are held in Ghana (INTERPOL, 2019). Information about match rates and the contribution or value of the held profiles is unknown. These accountability issues are partly due to the absence of oversight bodies for the use of forensic DNA in Ghana.

TABLE 4.1 Number and category of DNA profiles held in Ghana as of 2016

Profile category	*Number*
Reference	202
Crime scene	338
Missing person	3
Relatives of missing person	1
Unidentified human remains	1
Others	648
Total	1,193

Source: INTERPOL, 2016

Impact of DNA analysis in Ghana

The use of DNA evidence has been tested in Ghana's courts of law. Transcripts of some of those cases are available, and they provide an insight into forensic DNA typing in Ghana. We discuss these cases in this section as well as the portrayal of forensic DNA evidence in the media.

On 23 April 2002, the police received reports of an armed robbery at Lashibi, a suburb of Accra. The case involved the robbery of a vehicle and some electronic gadgets. Subsequent investigations by the police led to the arrest of four men, who were later convicted at the High Court, Fast Track Division, Accra. The evidence relied upon in the case included the recovery of the missing items from the homes of the alleged suspects and confession evidence. One of the convicted suspects, Mr Frimpong, appealed their conviction on the grounds of inappropriate evaluation of the circumstantial evidence by the court. Although this case did not feature DNA evidence, the Supreme Court, whilst commenting on the use of circumstantial evidence, classified DNA as acceptable evidence by the law courts (*Frimpong v The Republic*, 2012).

In *Asante v The Republic*, 2017, DNA evidence (paternity testing) was used to support the exoneration of an individual who was previously convicted by the High Court, Tamale, for an alleged defilement (i.e., sexual intercourse with a minor) in 2003. The initial conviction in 2005 was based on the account of the complainant (then a juvenile) and a positive pregnancy test. At the time of the trial, no DNA or scientific tests were carried out to determine the paternity of the child. A subsequent appeal of the conviction was dismissed in 2006 by the Court of Appeal, and in 2012, the Supreme Court granted leave for a second appeal. Following court approval in 2014, the FSL carried out a DNA test of the appellant and the child from the disputed pregnancy in 2015, which excluded the appellant as the father. Commenting on the use of DNA evidence in criminal investigations, the Supreme Court noted its high accuracy and reliability, as well as the potential to help resolve sexual offences (*Asante v The Republic*, 2017). This is largely consistent with comments made by the courts in other jurisdictions (National Research Council [US] Committee on DNA Forensic Science, 1996; *R v Bates*, 2006; *R v Doheny*, 1996). An issue that emerged from the media reports of the case was the delay in DNA testing and the lack of transparency in the test procedure. For example, Graphic Online reported that the prosecution team was not aware of the DNA tests (Bokpe and Akese, 2016).

Since the establishment of the FSL and the first uses of DNA evidence in court, the Ghanaian media has promoted the value of forensic science and reported real-life information about the use of forensics by the police (Akese, 2015; Daily Guide, 2011). Generally, the portrayal of the science of forensic DNA analysis in the Ghanaian media is positive, with several news reports highlighting its accuracy and objectivity and its use in other jurisdictions (GhanaWeb, 2019; Mensah, 2019; Savage, 2019). However, reports of police investigation issues and trends in criminal activities have resulted in calls by stakeholders for the effective utilisation and regulation of forensic services, including the adoption of an independent structure for the FSL to improve public confidence in the criminal justice system (Baneseh, 2018; Bokpe and Akese, 2016; Effah, 2019; Gadugah, 2015; Ghana News Agency, 2013; Mensah, 2016). These discussions have mainly occurred in the media due to the absence of an authoritative body or professional society on forensic science.

The Takoradi missing girls case illustrates some of the critical issues associated with forensic DNA practice in Ghana. The four young women disappeared in 2018 in the Western Region of Ghana, and it soon turned out that they were kidnapped. The unknown perpetrators demanded ransom for their release (Myjoyonline, 2019). Following a protracted investigation and intelligence challenges, the police discovered unidentified human remains at the residence of a key suspect in the case. In September 2019, based on DNA analyses, the police confirmed that the human remains belonged to the missing girls. The investigation received intense media and public attention, with demands for accountability and transparency from politicians and the families of the victims (Modern Ghana, 2019). Although the DNA evidence provided some form of closure in the case, there were controversies about the management of the investigation, the lack of details about the DNA analysis conducted by the FSL and the interpretation of the results (Ghana News Agency, 2019). For example, the police initially reported that they had found the whereabout of the missing persons, which was later revealed to be false, with the families accusing the police of deceit (Myjoyonline, 2019). The conflicting reports from the police subsequently damaged public trust in the integrity of the investigation. These issues highlight, among other things, the relevance of the principles of accountability and transparency in the governance of forensic science and DNA analysis in Ghana. In 2021, the two key suspects in the investigation were sentenced to death at the Sekondi High Court based on DNA and other circumstantial evidence (Aklorbortu and Dzodzegbe, 2021).

These case studies demonstrate how the absence of robust quality management systems and governance structure can lead to mistrust in police investigations and the criminal justice system. The available published evidence on the impact of forensic DNA analysis shows limited use of the technology to support the identification of suspects and conviction or exoneration of individuals. Whilst the evidence suggests the potential of DNA analysis in assisting the police to solve crime and missing person cases, as well as correcting miscarriages of justice, several legal, regulatory and quality issues need to be addressed. To fully understand the value and impact of the use of forensic DNA evidence in Ghana, there is a need for authorities to collect data on the number of cases involving DNA evidence and its impact on case outcomes and make this information available in the public domain. Currently, no such data is provided, and requests for information from authorities have been a challenge with several bureaucratic hurdles. The lack of data on the impact and effectiveness of forensic science, including the use of DNA evidence, is a contemporary issue that has also been raised in Europe and other jurisdictions (Amankwaa and McCartney, 2019; Toom, 2014; Toom, Granja, and Ludwig, 2019; Wiles, 2020).

Governance principles for the use of forensic DNA

In this last part, we discuss the key governance principles proposed in literature to govern the technolegal world of forensic DNA analysis and databases. The collection, retention and use of DNA for policing purposes is a widely debated topic in the socio-legal scholarship on forensic genetics (e.g., Machado and Silva, 2019; M'charek, 2008; Skinner and Wienroth, 2019; Toom, Granja, and Ludwig, 2019). This is mainly because DNA holds sensitive information about our health, predisposition to diseases, biological relationships and ancestry. For these reasons, countries that use DNA in criminal investigations adhere to various standards ensuring ethical and proportionate use by authorities (M'charek, 2008). These principles include viability, legitimacy and acceptability (McCartney, 2014; McCartney, Wilson, and Williams, 2011), as well as effectiveness, transparency and accountability (Toom, Granja, and Ludwig, 2019). We consider these principles in the context of DNA profiling and databasing in Ghana.

Viability is concerned with the quality, reliability and transparency of forensic information, such as matching DNA profiles, and the processes used to produce that information (McCartney, 2014; Wienroth, 2020). It is an important principle on which other concepts, such as acceptability and effectiveness, are established. The concept of viability involves using appropriate and proportionate procedures and meeting prescribed standards, which ensure the production of quality outcomes. Assessing viability in Ghana's case, it is important to evaluate whether forensic laboratories are meeting the required international standards and using the right protocols and procedures to prevent errors from being made to help ensure the credibility and reliability of results. Although the police FSL and other laboratories in the country provide DNA testing services for criminal and/or civil cases, these laboratories have not adopted international accreditation standards, such as ISO 17025 and ISO 17020 (Amankwaa et al., 2019). In addition, there are, as far as we know, no internationally accepted quality management systems ensuring that forensic labs are employing best practices and validated methods in crime scene investigations, laboratory analysis and the interpretation, evaluation and presentation of forensic evidence. The absence of robust technical guidance and standards to regulate the activities of the FSL and other forensic providers suggests that the reliability and the admissibility of forensic evidence may be questionable (i.e., subject to potential challenges in court). The operation of the FSL and forensic providers remains arcane, with limited published data on the validation of methods and procedures, competency requirements of personnel, record-keeping and disclosure requirements, environmental conditions and management of potential errors and contamination. The status of forensic science provision in Ghana means that miscarriages of justice involving problematic forensic evidence may be difficult to detect. The lack of operational transparency has raised concerns about trust in the police and the competence of the FSL to carry out DNA analysis (Agyei Annim, 2021). In this respect, the current status of forensic DNA typing in Ghana demonstrates some similarities to the introduction of this technology in the US in the late 1980s and early 1990s (Lander, 1989; Lynch et al., 2008; M'charek, 2000).

Legitimacy is another key principle that encompasses the justifications for and the establishment of appropriate legal boundaries on the creation, retention, use and destruction of forensic DNA information/material (McCartney, 2014; Wienroth, 2020). It is based on the concepts of justification, lawfulness, the ethical and proportionate application of any interference with rights and 'doing the right thing'. In this context, where human subjects are involved, legitimacy spells out laws that ensure the right treatment is meted out to individuals, and their rights are protected in every interaction. As the FSL and other forensic providers gather, retain, utilise and share forensic data/material, it is crucial to assess the legitimacy of their activities under existing law and within the wider societal context. In Ghana, specific laws that regulate forensic activities are currently not in existence, although there are some general regulations, including the Data Protection Act 2012, which classifies DNA as special personal data. A major legal gap is the absence of a clear statutory basis for the acquisition, inclusion, use, retention and destruction of forensic DNA samples and profiles (Amankwaa et al., 2019). Further research investigation is required to determine the extent of the impact of this gap on the criminal investigation process and challenges to the admissibility of DNA evidence. In many jurisdictions, the introduction and further development of forensic DNA practices have followed a common trajectory (Williams and Johnson, 2008), in which DNA typing is first used on a case-by-case basis in violent crimes and is later routinely deployed in other cases, like volume crimes. Cases in which the Ghana police have successfully used DNA evidence demonstrate that Ghana's forensic DNA practices are still in the early phase of such common trajectory (Aklorbortu and Dzodzegbe, 2021; *Asante v The Republic*, 2017).

A third principle, **acceptability**, is based on considerations of democratic concepts that require the input of citizens in establishing the laws of the country (McCartney, 2014; McCartney, Wilson, and Williams, 2011). Acceptability is also based on trust in a state and its institutions and trust in technology. As mentioned earlier, acceptability builds on viability. Individuals tend to trust and believe in a system that can evidence itself to be credible, reliable and efficient. In Ghana, for instance, the proliferation of video surveillance over the years has been massive. It is common to come across surveillance cameras installed in various business facilities and some homes, even when the services of security personnel are employed. This is mainly because people trust in the power of technology to help prevent and combat crime by assisting the police in apprehending perpetrators. In 2019, the Ghana Police Service started the installation of about 1,000 surveillance cameras across the country, which many citizens allegedly consider as a positive step towards curbing crime (Salia, 2019). Just as with video surveillance, there appears to be public trust in the potential of forensic DNA to help solve crimes, although there are concerns about the quality assurance processes of the FSL (Agyei Annim, 2021; Savage, 2019). While the use of forensic DNA has been in the national news regarding its value in specific cases, there has been little or no public debate on the establishment of an appropriate legal framework.

In addition to viability, legitimacy and acceptability, effectiveness and accountability are critical to the implementation of forensic DNA (Amankwaa, 2019; Amankwaa and McCartney, 2019; IAG, 2018, Toom, Granja, and Ludwig, 2019). **Effectiveness** involves the evaluation of the actual outcomes of a system in terms of whether it meets certain pre-defined expectations (Amankwaa and McCartney, 2019). In measuring the effectiveness of forensic DNA and databases, Amankwaa and McCartney (2019) have identified seven indicators: 'crime-solving capacity, incapacitation effect, deterrence effect, privacy protection, legitimacy, implementation efficiency and implementation cost'. There is presently no available data to systematically assess the effectiveness of the use of forensic DNA in Ghana using these parameters. Nevertheless, there are a few criminal cases in which DNA evidence played a critical role in apprehending offenders or exonerating an individual who had been wrongfully convicted (Aklorbortu and Dzodzegbe, 2021; *Asante v The Republic*, 2017). As a result, there are some indicators of the potential crime-solving capacity and incapacitation effects of the use of forensic DNA.

Lastly, **accountability** is essential to ensure that the use of forensic DNA is compatible with the law and compliant with established standards. This principle is mainly applied through the establishment of independent governance or regulatory bodies, which serve as checks and balances vis-à-vis forensic activities by service providers and the police (Amankwaa, 2019; FIND Strategy Board, 2020; IAG, 2018; Tully, 2021). Yet, no independent forensic science regulator or national DNA database strategy board, which provides codes of practice and conduct or to which forensic science providers are accountable, such as are in place in England and Wales, have been established in Ghana (FIND Strategy Board, 2020). This conundrum, therefore, explains the ongoing public mistrust of the work of the police FSL (Agyei Annim, 2021; Amankwaa et al., 2019).

In summary, although the FSL has commenced the collection, retention and use of DNA records, the application of the five key principles outlined here is inadequate in the regulation of DNA profiling and databasing. Our analysis demonstrates that the emerging technolegal world in Ghana is unregulated and tenuous. As stated earlier, there has been no public discourse or initiative to establish a legal and operational framework for forensic DNA profiling and databasing. The laboratories of the FSL are yet to be accredited against international standards, such as ISO 17025 (ISO, 2017), to assure the quality of DNA practices (ENFSI DNA Working Group, 2017). Further, there is presently no specific statutory basis for DNA profiling and databasing for policing purposes, and research on the impact and effectiveness of forensic science is lacking. To improve the status of DNA profiling and databasing in Ghana, the FSL and other forensic providers should adopt the specific recommendations of international agencies, such as the ENFSI DNA Working Group, which align with the five key principles for the governance of forensic DNA information.

Conclusion

There is a growing acceptance of forensic DNA analysis and databasing in the justice system of Ghana since the EU-Ghana Police Project in 2011. This investigative tool has contributed to a few criminal investigations and assisted in correcting miscarriages of justice. Available reviews on forensic science in Ghana suggest a wide scope and potential for the application of DNA profiling and databasing. However, there are critical gaps in the governance of this technolegal world, including issues concerning the accreditation of the FSL, validation of DNA analysis processes and retention of DNA data, establishment of a core standard set STR loci and an allele frequency database and the absence of a dedicated legislative framework and independent regulatory bodies.

To develop forensic DNA practice in Ghana, a national policy strategy should be established by relevant stakeholders to improve the infrastructure for DNA profiling and databasing. This should include accreditation of DNA laboratories and research into the allele and haplotype frequencies for the different DNA polymorphisms for the major ethnic groups in Ghana. The availability of these population frequency databases will ensure the robust and transparent interpretation and evaluation of forensic DNA evidence. Further, there should be a national agreement on the core loci for forensic casework to ensure consistency in practice and support for the establishment of a national forensic DNA database and the transnational exchange of DNA information.

Another critical area for improvement is the incorporation of the five governance principles of DNA profiling and databasing in specific legislation. The adoption of a specific legal/governance framework from the onset of DNA profiling and databasing is highly recommended in the literature (M'charek, Hagendijk, and de Vries, 2013; National Research Council [US] Committee on DNA Technology in Forensic Science, 1992; Toom, 2012). This approach is considered effective in ensuring that the ethical costs of the use of forensic DNA and databases, such as interference with privacy rights, are carefully weighed and protected by authorities. Further, this model is thought to be more efficient and cost effective in assisting law enforcement authorities in achieving the public security objectives of the criminal justice system.

Notes

1 See Ghana Police. (n.d.). Forensic Science Laboratory. Retrieved 14 August 2020, from https://police.gov.gh/en/index.php/forensic-science-laboratory-fsl/.
2 The following keywords and string search were used for the literature search: ("forensic DNA analysis" OR "STR typing" OR "DNA fingerprinting" OR "DNA profile" OR "DNA profiling" OR "DNA database") AND Ghana.

References

ACP, European Community, 2000. PARTNERSHIP AGREEMENT between the members of the African, Caribbean and pacific group of states of the one part, and the

European community and its member states, of the other part, signed in Cotonou on 23 June 2000. *Official Journal of the European Communities L* 317, 3–64.

Afrifah, K.A., Badu-Boateng, A., Antwi-Akomeah, S., Motey, E., Boampong, E., Twumasi, P., Sampene, P., Donkor, A., 2020. Forensic identification of missing persons using DNA from surviving relatives and femur bone retrieved from salty environment. *Journal of Forensic Science and Medicine* 6, 40. https://doi.org/10.4103/jfsm.jfsm_48_19.

Agyei Annim, A., 2021. Families of murdered Takoradi girls demand DNA results, state burial. *Citinewsroom—Comprehensive News in Ghana*. https://citinewsroom.com/2021/03/families-of-murdered-takoradi-girls-demand-dna-results-state-burial/ (accessed 3.30.21).

Akese, E., 2015. DNA proves man innocent: After 10 years in prison [WWW Document]. Graphic Online. www.graphic.com.gh/news/general-news/dna-proves-man-innocent-after-10-years-in-prison.html (accessed 6.12.19).

Aklorbortu, D.K., Dzodzegbe, A., 2021. Takoradi missing girls judgment: 2 Kidnappers to die. Graphic Online.

Amankwaa, A.O., 2019. Towards a reformed policy for immigrant DNA tests, a commentary. *Journal of Forensic and Legal Medicine* 66, 117–119. https://doi.org/10.1016/j.jflm.2019.06.016.

Amankwaa, A.O., McCartney, C., 2019. The effectiveness of the UK national DNA database. *Forensic Science International: Synergy* 1, 45–55. https://doi.org/10.1016/j.fsisyn.2019.03.004.

Amankwaa, A.O., Nsiah Amoako, E., Bonsu, D.O.M., Banyeh, M., 2019. Forensic science in Ghana: A review. *Forensic Science International: Synergy* 1, 151–160. https://doi.org/10.1016/j.fsisyn.2019.07.008.

Asante v The Republic, 2017.

Badu-Boateng, A., Twumasi, P., Salifu, S.P., Afrifah, K.A., 2018. A comparative study of different laboratory storage conditions for enhanced DNA analysis of crime scene soil-blood mixed sample. *Forensic Science International* 292, 97–109. https://doi.org/10.1016/j.forsciint.2018.09.007.

Baneseh, M.A., 2018. Police hospital, crime lab give conflicting results in Lebanese rape case [WWW Document]. Graphic Online. www.graphic.com.gh/news/general-news/police-hospital-crime-lab-give-conflicting-results-in-lebanese-rape-case.html (accessed 6.12.19).

BBC News, 2015. New DNA facility gives hope of "cold case" justice in Scotland. *BBC News*.

Bokpe, S.J., Akese, E., 2016. Supreme court scolds A-G Dept over DNA results [WWW Document]. Graphic Online. www.graphic.com.gh/news/general-news/supreme-court-scolds-a-g-dept-over-dna-results.html (accessed 6.12.19).

Butler, J., 2012. *Advanced Topics in Forensic DNA Typing: Methodology*. Elsevier, London.

CIA, 2020. The world factbook: Africa: Ghana [WWW Document]. *Central Intelligence Agency*. www.cia.gov/library/publications/the-world-factbook/geos/gh.html (accessed 1.28.20).

Daily Guide, 2011. Police gets forensic Lab [WWW Document]. www.ghanaweb.com/GhanaHomePage/NewsArchive/Police-Gets-Forensic-Lab-226024 (accessed 6.18.19).

DEUG, 2012. Quarterly Newsletter of the Delegation of the European Union to Ghana. *Delegation of the European Union to Ghana* 1–9.

Effah, S., 2019. Forensic scientists push for independent forensic lab in Ghana. *3news*. https://3news.com/forensic-scientists-push-for-independent-forensic-lab-in-ghana/ (accessed 3.31.21).

ENFSI DNA Working Group, 2017. *DNA Database Management Review and Recommendations*. ENFSI, Wiesbaden.

Fendt, L., Röck, A., Zimmermann, B., Bodner, M., Thye, T., Tschentscher, F., Owusu-Dabo, E., Göbel, T.M.K., Schneider, P.M., Parson, W., 2012. MtDNA diversity of Ghana: A forensic and phylogeographic view. *Forensic Science International: Genetics* 6, 244–249. https://doi.org/10.1016/j.fsigen.2011.05.011.

FIND Strategy Board, 2019. *National DNA Database: Annual Report, 2017 to 2018.* Forensic Information Database Strategy Board, London.

FIND Strategy Board, 2020. *National DNA Database Strategy Board Biennial Report 2018–2020.* Forensic Information Database Strategy Board, London.

Frimpong v The Republic, 2012.

Gadugah, N., 2015. June 3 disaster caused by cigarette smoker-Committee—MyJoyOnline.com [WWW Document]. *Myjoyonline.* www.myjoyonline.com/news/june-3-disaster-caused-by-cigarette-smoker-committee/ (accessed 7.2.19).

Ghana News Agency, 2013. US experts arrive in Kumasi to investigate fire outbreak. *Ghana Business News.* www.ghanabusinessnews.com/2013/06/18/us-experts-arrive-in-kumasi-to-investigate-fire-outbreak/ (accessed 7.4.19).

Ghana News Agency, 2019. Families of Takoradi kidnapped girls express disappointment at mode of communication [WWW Document]. *GhanaWeb.* www.ghanaweb.com/GhanaHomePage/NewsArchive/Families-of-Takoradi-kidnapped-girls-express-disappointment-at-mode-of-communication-781827 (accessed 12.17.21).

GhanaWeb, 2019. Takoradi missing girls were killed nine months ago—DNA test report reveals [WWW Document]. *Ghana Web.* www.ghanaweb.com/GhanaHomePage/NewsArchive/Takoradi-missing-girls-were-killed-nine-months-ago-DNA-test-report-reveals-782505 (accessed 8.27.20).

IAG, 2018. *Report of the Independent Advisory Group on the Use of Biometric Data in Scotland.* Scottish Government, Edinburgh.

ILAC, 2014. *Modules in a Forensic Science Process, ILAC G19:08/2014.* International Laboratory Accreditation Cooperation, Silverwater, Australia.

INTERPOL, 2016. *Global DNA Profiling Survey Results 2016.* INTERPOL, Lyon, France.

INTERPOL, 2019. *INTERPOL Global DNA Profiling Survey Results 2019.* INTERPOL, Lyon, France.

ISO, 2017. *ISO/IEC 17025:2017: General Requirements for the Competence of Testing and Calibration Laboratories.* ISO, Geneva.

Kofi, A.E., Hakim, H.M., Khan, H.O., Ismail, S.A., Ghansah, A., Haslindawaty, A.R.N., Shamsuddin, S., Aziz, M.Y., Chambers, G.K., Edinur, H.A., 2020. Population dataset for 21 simple tandem repeat loci in the Akan population of Ghana. *Data in Brief* 31, 105746. https://doi.org/10.1016/j.dib.2020.105746.

Koomson, A., Gaisie, E.K., Ayitey, D.T., Antiaye, N.A., 2019. Forensic science: A unique interdisciplinary tool helping to combat crimes in Africa, a review. *FSSGH Newsletter* 2, 7.

Lander, E.S., 1989. DNA fingerprinting on trial. *Nature* 339, 501–505. https://doi.org/10.1038/339501a0.

Lynch, M., Cole, S., Mcnally, R., Jordan, K., 2008. *Truth Machine: The Contentious History of DNA Fingerprinting.* University of Chicago Press, Chicago.

Machado, H., Silva, S., 2019. What influences public views on forensic DNA testing in the criminal field? A scoping review of quantitative evidence. *Human Genomics* 13, 23. https://doi.org/10.1186/s40246-019-0207-5.

McCartney, C.I., 2014. Forensic data exchange: Ensuring integrity. *Australian Journal of Forensic Sciences* 47, 36–48. https://doi.org/10.1080/00450618.2014.906654.

McCartney, C.I., Wilson, T.J., Williams, R., 2011. Transnational exchange of forensic DNA: Viability, legitimacy, and acceptability. *European Journal on Criminal Policy and Research* 17, 305–322. https://doi.org/10.1007/s10610-011-9154-y.

M'charek, A., 2000. Technologies of population: Forensic DNA testing practices and the making of differences and similarities. *Configurations* 8, 121–159. https://doi.org/10.1353/con.2000.0005.

M'charek, A., 2008. Silent witness, articulate collective: DNA evidence and the inference of visible traits. *Bioethics* 22, 519–528. https://doi.org/10.1111/j.1467-8519.2008.00699.x.

M'charek, A., Hagendijk, R., de Vries, W., 2013. Equal before the law: On the machinery of sameness in forensic DNA practice. *Science, Technology, & Human Values* 38, 542–565.
Mensah, L., 2016. Let's invest more resources in the security of Ghana [WWW Document]. *Centre for International Education*. https://cie.ucc.edu.gh/news/let%E2%80%99s-invest-more-resources-security-ghana (accessed 8.27.20).
Mensah, M., 2019. Takoradi missing girls dead; DNA tests confirm [WWW Document]. Graphic Online. www.graphic.com.gh/news/general-news/takoradi-missing-girls-dead-dna-tests-confirm.html (accessed 9.18.19).
Modern Ghana, 2019. Missing Takoradi girls: The story so far [WWW Document]. *Modern Ghana*. www.modernghana.com/news/948778/missing-takoradi-girls-the-story-so.html (accessed 12.17.21).
MOFEP, 2009a. *Support to the Ghana Police Service: Tender Dossier: Equipment for the Forensic Science Laboratory of the Criminal Investigation Department (EuropeAid/129238/D/SUP/EU)*. Ministry of Finance & Economic Planning, Accra.
MOFEP, 2009b. *Support to the Ghana Police Service: Tender Dossier: Supply of Vehicles for Equipping the Forensic Science Laboratory of the Criminal Investigation Department (EuropeAid/129–498/M/SUP/GH)*. Ministry of Finance & Economic Planning, Accra.
MOFEP, 2010. *Support to the Ghana Police Service: Tender Dossier: Supply of Equipment and Software for the Forensic Science Laboratory of the Criminal Investigation Department (EuropeAid/130125/M/SUP/GH)*. Ministry of Finance & Economic Planning, Accra.
Murphy, E., 2013. Physician heal thyself: Whither the police and prosecutor in the tale of forensic science gone wrong? *Texas Law Review* 91, 101–113.
Myjoyonline, 2019. Timeline of missing Takoradi girls' saga: The story so far [WWW Document]. URL www.myjoyonline.com/news/timeline-of-missing-takoradi-girls-saga-the-story-so-far/ (accessed 9.18.19).
National Research Council (US) Committee on DNA Forensic Science, 1996. *DNA Evidence in the Legal System, The Evaluation of Forensic DNA Evidence*. National Academies Press (US), Washington, DC.
National Research Council (US) Committee on DNA Technology in Forensic Science, 1992. *DNA Technology in Forensic Science*. National Academies Press (US), Washington, DC.
Parliament of the Republic of Ghana, 2012. Data Protection Act.
Poetsch, M., Ergin, Z., Bayer, K., El-Mostaqim, D., Rakotomavo, N., Browne, E.N.L., Timmann, C., Horstmann, R.D., Schwark, T., von Wurmb-Schwark, N., 2011. The new Powerplex ESX17 and ESI17 kits in paternity and maternity analyses involving people from Africa—including allele frequencies for three African populations. *International Journal of Legal Medicine* 125, 149–154. https://doi.org/10.1007/s00414-010-0502-0.
Prainsack, B., Toom, V., 2013. Performing the Union: The Prüm Decision and the European dream. *Studies in History and Philosophy of Science Part C: Studies in History and Philosophy of Biological and Biomedical Sciences* 44, 71–79. https://doi.org/10.1016/j.shpsc.2012.09.009.
R v Bates, 2006.
R v Doheny, 1996.
Salia, A.K., 2019. Police deploy 1000 cameras across Ghana for monitoring. Graphic Online.
Sanchez-Faddeev, H., Pijpe, J., van Bodegom, D., van der Hulle, T., van der Gaag, K.J., Eriksson, U.K., Spear, T., Westendorp, R.G.J., de Knijff, P., 2013a. Ancestral stories of Ghanaian Bimoba reflect millennia-old genetic lineages. *PLoS One* 8. https://doi.org/10.1371/journal.pone.0065690.
Sanchez-Faddeev, H., Pijpe, J., van der Hulle, T., Meij, H.J., van der Gaag, K.J., Slagboom, P.E., Westendorp, R.G.J., de Knijff, P., 2013b. The influence of clan structure on the genetic variation in a single Ghanaian village. *European Journal of Human Genetics* 21, 1134–1139. https://doi.org/10.1038/ejhg.2013.12.

Savage, J., 2019. When will Ghana embrace DNA-forensic method for crime investigation? [WWW Document]. *Modern Ghana*. www.modernghana.com/news/909399/when-will-ghana-embrace-dna-forensic-method-for.html (accessed 8.27.20).

Science and Technology Committee, 2005. *Forensic Science on Trial: Seventh Report of Session 2004–05* (No. HC 96-I). House of Commons, London.

Skinner, D., Wienroth, M., 2019. Was this an ending? The destruction of samples and deletion of records from the UK Police National DNA Database. *BJHS Themes* 1–23. https://doi.org/10.1017/bjt.2019.7.

Toom, V., 2012. Forensic DNA databases in England and the Netherlands: Governance, structure and performance compared. *New Genetics and Society* 31, 311–322. https://doi.org/10.1080/14636778.2012.687133.

Toom, V., 2014. Trumping communitarianism: Crime control and forensic DNA typing and databasing in Singapore. *Easts* 8, 273–296. https://doi.org/10.1215/18752160-2416805.

Toom, V., Granja, R., Ludwig, A., 2019. The Prüm Decisions as an aspirational regime: Reviewing a decade of cross-border exchange and comparison of forensic DNA data. *Forensic Science International: Genetics* 41, 50–57. https://doi.org/10.1016/j.fsigen.2019.03.023.

Tully, G., 2021. *Forensic Science Regulator: Annual Report 2019–2020*. Forensic Science Regulator, Birmingham.

Wallace, H., 2006. The UK national DNA database: Balancing crime detection, human rights and privacy. *EMBO Reports* 7, S26–S30. https://doi.org/10.1038/sj.embor.7400727.

Wang, D.Y., Gopinath, S., Lagacé, R.E., Norona, W., Hennessy, L.K., Short, M.L., Mulero, J.J., 2015. Developmental validation of the GlobalFiler® Express PCR Amplification Kit: A 6-dye multiplex assay for the direct amplification of reference samples. *Forensic Science International: Genetics* 19, 148–155. https://doi.org/10.1016/j.fsigen.2015.07.013.

Wepeba, P.-P., Iyengar, A., Goodwin, W., 2019. Heterozygous 21 STR loci and triplet alleles observed in population genetic analysis of the GlobalFiler STR loci in the Ghanaian population. *Forensic Science International: Genetics Supplement Series, The 28th Congress of the International Society for Forensic Genetics* 7, 753–754. https://doi.org/10.1016/j.fsigss.2019.10.164.

Wienroth, M., 2020. Value beyond scientific validity: Let's RULE (Reliability, Utility, LEgitimacy). *Journal of Responsible Innovation* 7, 92–103. https://doi.org/10.1080/23299460.2020.1835152.

Wiles, P., 2020. *Annual Report 2019: Commissioner for the Retention and Use of Biometric Material*. Office of the Biometrics Commissioner, London.

Williams, R., Johnson, P., 2008. *Genetic policing: The uses of DNA in police investigations*. Cullompton, Devon: Willan Publishing.

5

LEGISLATING FORENSIC GENETICS IN SOUTH AFRICA

Science, justice and the occlusion of race in postapartheid DNA databasing

Noah Tamarkin

Introduction

South Africa was among the first countries in the world to have a national criminal DNA database, but for many years its small scale and lack of governing legislation limited its significance and functionality.[1] This changed in 2015, when South Africa began to implement a law that had been passed two years before, the Criminal Law (Forensic Procedures) Amendment Act. Known as the DNA Act, this law laid the foundation for formalising a national forensic DNA database. In doing so, South Africa joined and contributed to the well-established and ever-growing global proliferation of national DNA databases, even while it also ushered in a new, specifically South African technolegal world of forensic genetics.

This chapter provides an empirical account of this South African technolegal world in formation. It considers approximately fifteen years of parliamentary records that discuss forensic genetics, from the late 1990s, when a small database existed but was unregulated, through 2009, when what became the DNA Act was first discussed but could not be quickly passed, to 2015, when the finalised legislation began to be implemented. At its largest in 2018, the DNA database that this law legitimised contained over 1.5 million profiles, though more recently, as expungements of those arrested but exonerated have outpaced additions of newly arrested and convicted people, it has stabilised at around 750,000 profiles.[2]

Today's political terrain is not the same as that of South Africa's earliest DNA database in the 1990s: during this time, the excitement and possibility of the early years of postapartheid transformation have given way to new forms of political and social contestation. This chapter aims to account for how legislating forensic genetics intersected with these shifts, with particular attunement to reverberations of South Africa's colonial and apartheid histories. It asks, what ideas and forms of race, science and justice were called into question and stabilised as DNA databasing

DOI: 10.4324/9780429322358-7

in South Africa moved from technological potentiality to legally mandated policing policy?

Three arguments about the role of race in forensic genetics anchor this chapter. First, I argue that because racial categorisation determined one's access to rights and citizenship in colonial and apartheid South Africa, race must be viewed as a foundational concern when considering legislation that fundamentally reconfigures rights by establishing instances when they can be violated (Toom, 2012a). Second, I show that the changing status of race vis-à-vis what became the DNA Act can't be understood apart from considering how science and justice were also articulated to the same phenomena and how those articulations also shifted over time. Third, I argue that the shifting meaning and significance of race in South Africa's DNA legislation process offers an opening to analyse how race, racialisation and racism articulate to forensic genetics practices more broadly, even – perhaps especially – when such articulations are not self-evident or easily identifiable.

This chapter builds on the comparative work of STS scholars who have considered implications of national DNA databases and their governance, primarily in the UK, throughout the European Union and in the United States (Amelung et al., 2021; Hindmarsh and Prainsack, 2010; Krimsky and Simoncelli, 2011; Machado and Granja, 2020; Machado et al., 2019; Toom, 2012b; Vailly and Bouagga, 2019; Wienroth, 2018). National DNA databases have different rationales, different modes of contestation, different resonances and different implications. Nevertheless, situated choices made in organising, governing and restricting national DNA databases reverberate internationally. South African forensic genetics, therefore, provides an important case study for two reasons. First, the country's colonial and apartheid history and postapartheid political imperatives make especially clear the connections between race and forensic genetic technologies and practices. Second, the specificities of South African forensic genetic legislation are likely to be an important point of reference and influence going forward, especially if other postcolonial states that are increasingly establishing or considering national forensic genetic investments seek models outside the global North.

It can seem like the expansion of national forensic DNA databases throughout the world is inevitable, but in any given instance, such a database comes from particular convergences of interests, only some of which correspond to the international zeitgeist of forensic genetics (Hindmarsh and Prainsack, 2010; Skinner and Wienroth, 2019). This chapter excavates how South Africa's national forensic DNA database was established legislatively. It considers parliamentary records that discuss forensic genetics from the late 1990s, when a small database existed but was unregulated, through 2009, when draft legislation was discussed but not adopted, to 2015, when the finalised legislation, passed in 2013, began to be implemented. It aims to account for how legislating forensic genetics intersected with social and political shifts from the 1990s to 2015, with particular attunement to reverberations of South Africa's colonial and apartheid histories. This chapter argues that the shifting meaning and significance of race in South Africa's DNA legislation process offers an opening to analyse how race, racialisation and racism articulate to forensic genetics practices more broadly; that racialised logics of policing and securitisation are at the core of forensic DNA databases; and that a significant part of the labour and contingency of forensic DNA databases lies in the reconfiguration and occlusion of race.

Biometric histories, racialised carcerality: Colonial and apartheid South African policing

DNA databases and the forensic genetics practices that they make possible are part of deeper biometric histories: they build on and carry forward earlier epistemologies and practices that developed through fingerprinting technologies in the 19th century (Breckenridge, 2014; Cole and Lynch, 2010, p. 108; Lynch et al., 2008). Before there was a South African state, the Transvaal police force relied on fingerprinting as a way to simultaneously identify the black laborers who worked Johannesburg's gold mines and criminalise them for violating pass laws, which dictated under what conditions they could be present in the city and mandated incarceration as a penalty for job abandonment (Breckenridge, 2014, pp. 72–74). This was known in South Africa as 'influx control', and is the core of what Mahmood Mamdani has called the bifurcated state, in which white citizens had rights and black subjects, who were considered legitimate residents of rural 'native reserves' and only temporary laborers in cities, did not (Mamdani, 1996). When South Africa became a unified state in 1910, it inherited the Transvaal police force along with three other colonial police forces and, in 1913, merged them into one to found the South African Police (SAP) (Brewer, 1994; Hornberger, 2011). South Africa's racial carcerality (Davis, 2003; Gilmore, 2007) continued throughout the first half of the 20th century. It reached new extremes, including the total loss of citizenship for Black South Africans, under apartheid laws put in place by the Afrikaaner nationalist National Party, in power from 1948 until 1994 (Beinart and Dubow, 1995; Dubow, 2014; Posel, 1991).[3]

The goals of apartheid policing were twofold: to intensify racialised policing and to criminalise political dissent, in part by expanding police power (Hornberger, 2011). The original Criminal Procedure Act of 1955 – the basis of the 1977 amended law that would be further amended in 2013 to account for police collection of DNA and a national DNA database – emerged in this context. Along with the Police Act of 1958, the Criminal Procedure Act of 1955 solidified state power at the expense of citizens' and subjects' rights and lives and allowed for things like warrantless searches and extended detention without conviction (Hornberger, 2011, p. 24). Under apartheid, policing became thoroughly securitised and militarised: security branch police infiltrated political organisations and underground movements, and they and public order police murdered activists and those perceived to be activists – including youth – with impunity.

In the 1990s and early 2000s transition to a postapartheid democratic South Africa, policing became a key site through which to work out in practice the newly open questions about the meaning of citizenship, criminality and race. The South

African Police, rebranded as the South African Police Service (SAPS) was high on the list of necessary institutional transformation, along with an entirely new constitution, new laws built around the idea of human rights and the complete repeal of laws designed to enforce apartheid. Having inherited a biometric state built on racial carcerality, the postapartheid ANC government was tasked with demilitarising the police, divesting from racial policing and scaling back the role of policing in the majority of peoples' interactions with the new, officially non-racial, democratic postapartheid state. All these goals were at odds with investing in new technologies at the intersection of biometrics and policing because such technologies evoked racialised carceral histories while expanding police power. So while forensic genetics had entered into postapartheid Truth and Reconciliation–related efforts to identify victims of security forces (Aronson, 2012), advocating for the biometric policing project of a national forensic DNA database would have been a surprising direction for the 1990s ANC postapartheid government to take. Meanwhile, forensic genetic technologies were being rapidly developed and adopted as a component of law enforcement practice around the world.

Making the case for a national DNA database: SAPS and parliament 1999–2001

Though it seemed an unlikely investment, postapartheid intersections between policing and forensic genetics had some momentum on their side. When the SAP Forensic Science Laboratory (FSL) was formed in 1971, biology was one of its founding units, and throughout the 1990s, the FSL worked with DNA in some sexual assault and child protection cases involving known suspects.[4] By 1999, the SAPS forensic science division had amassed what they called a DNA criminal intelligence database, or DCID, from the relatively small number of samples that they worked with, and they were hoping to expand it.

While in other countries like the UK, where DNA database expansion preceded regulation on the grounds that what was not explicitly prohibited was therefore possible, in 1990s South Africa, regulation was imagined as potentially productive of new practices while lack of regulation was imagined as prohibitive.[5] This was because the SAPS leadership was fully invested in a rehabilitative project of becoming known as protectors rather than violators of human rights (Hornberger, 2011). In the absence of justifying legislation, there was a risk that collecting, storing and comparing DNA might violate human rights to privacy and dignity, rights that were now central to the postapartheid constitution. Undermining these rights would undermine the SAPS fervently sought postapartheid legitimacy.[6]

So it was that in November of 1999, SAPS Forensic Sciences Division leadership made their case to the South African Parliament's Safety and Security Portfolio Committee for legislation that would 'give effect to the DCID'.[7] In their presentation, they emphasised the power of DNA to combat crime both nationally and across borders and the ways that the DCID, along with other functions of the Forensic Sciences Division, addressed the current national policing priorities and objectives and, in particular, the national effort to combat violence against women and children. In short, they made the case that a national investment in forensic genetics was an investment in a competent and protective state and, therefore, an investment in the nation's citizens: citizens to be protected were implicitly women and children, and the invocation of cross-border crime evoked the postapartheid iteration of influx control without naming it directly.

SAPS leaders were not operating in a vacuum. In these early postapartheid years, a small number of parliamentarians had raised questions about whether DNA testing, particularly as a means to determine paternity in rape cases that resulted in pregnancy, was part of South Africa's policing capacity and, if so, if all South Africans equally had access to this technology.[8] The two issues highlighted in this line of questioning – the need to address rape as a national crisis and the need to make racial, socio-economic and regional equity a national priority – set the stage for parliamentary interest in forensic genetics as something that had the potential to align with postapartheid transformative efforts. Here, the imagined potential of forensic genetics was contingent on the successful transformation of a reimagined police service that would be protective rather than oppressive, with full protective access to those who had been criminalised through apartheid policing priorities or subject to the separate Bantustan policing in rural areas.

At this early stage, the SAPS Forensic Science Division leadership, aided by sympathetic politicians, made the case for their potential to speak to national anxieties about transforming police from agents of racialised state violence to effective protectors of all South Africans. The questions they had to address were from politicians invested in racial justice, initially in the form of equal access to state services and, increasingly, as parliamentary discussions about forensic genetics advanced, in the form of affirmative action for diversifying SAPS employment.

Both the advocates of forensic genetics and their critics, in effect, then framed forensic genetics in terms of its capacity to work against inherited state racism: it would be part of transforming the police, not part of extending policing's racist legacies. However, at no point in the discussion was the DCID or the legislation that SAPS asked for either rejected or endorsed, nor were its potential racialised impacts raised. This was rather a case of the SAPS working to demonstrate that legislative and financial investments in forensic genetics aligned with national transformation of policing while politicians questioned whether one of the least racially diverse police divisions had yet proven itself sufficiently transformed to warrant such investments. Further, the focus on funding effectively turned conversations about racial justice and forensic genetics from equal access to police services towards equal access to police jobs for non-white South Africans.

With the fundamental need for a national DNA database unquestioned, some parliamentarians began to imagine how such a thing might be legislatively justified and to which other existing concerns it might provide a solution. In February 2000, just a few months after the SAPS made their case for a DCID, parliamentarians, notably including some from the dominant ANC party, considered including a provision in the Firearms Control Bill that would allow the collection of DNA samples from people who were found to be in the vicinity of illegal weapons. They noted that this would clearly violate the individual human rights of suspects, but they also opined that the constitutional provision that allowed for violation of some individual human rights if there was a good reason could be applied here.[9] Later that year, in a different committee meeting, a politician from the New National Party, which was the short-lived postapartheid successor to the National Party (the party behind the apartheid system), advocated for establishing a database of sex crime offenders that would include DNA samples.[10]

A population whose rights could be violated in the name of safety and security was beginning to take shape as SAPS and parliamentarians distanced themselves from state power as necessarily oppressive to instead reimagine it, in their hands, as benevolent and protective. That same year, some of the funding that SAPS officials sought also arrived: at some point in 2000, still lacking legislation to legitimise the DCID, the SAPS secured 120 million rand in foreign aid from the EU to establish the DNA database and train analysts to work with it. This was nearly half the foreign aid earmarked for bolstering the SAPS's ongoing transformative efforts that South Africa received that year.[11] EU funding lent further credibility to SAPS's leaders' and their parliamentarian allies' ideas of safety and security. A cross-party (ANC and NNP) and international (EU funding) consensus was thus beginning to build that investing in state capacity for forensic genetics would further investments in a new, transformed, safer South Africa, all of which began to shift a South African forensic DNA database from surprising to inevitable.

Legislating science, exceptionalising forensics: 2003–2008

By 2003, DNA enthusiasm had hit a snag. Just as had been feared several years earlier, in the absence of explicit regulations for how DNA could be collected and stored, genetic evidence had been facing court challenges regarding its admissibility. This had two opposite effects: it increased motivation to create legislation that might make DNA evidence effective in prosecutions, but it also slowed the process down because of a need to research how admissibility might be achieved.[12]

Meanwhile, two other bills were under discussion that, in different ways, illustrate the terms through which forensic genetics could at that time be advanced or contested. These two bills, the Natural Scientific Professions Bill, introduced in 2003, and the Refugees Amendment Bill, introduced in 2008, together bring into focus intersections of law and science as they were playing out in postapartheid South Africa and specifically questions about the role of the state in validating scientific authority, whether everyone should be equally bound by legislative mandates, and to whom constitutional protections and responsibilities extended. Together, these bills point to how legal definitions of science (here, forensic science and biometrics) clarified national discussions about race (affirmative action and anti-African xenophobia, respectively).

The Natural Scientific Professions Bill was introduced in 2003 with the goal of establishing a South African Council for Natural Scientific Professions that would register and regulate scientists working in South Africa; the idea was to ensure the qualifications and conduct of working scientists and to afford them both national and international legitimacy.[13] The bill stipulated that scientists must register with the council in order to practice science or work in any form of scientific consultancy, and it called for three categories of registered scientists: professional, candidate and certified natural scientists, such that only professionals with corresponding postgraduate degrees would be able to practice without supervision.

SAPS leadership objected because with no forensic science degree programs in South Africa, there could be no registered forensic scientists and therefore no legitimately operating SAPS forensic science laboratory unless the FSL budgets were expanded to accredit their in-house training; funds would also be needed for all their employees to register as scientists once their training was completed. Their proposed solution was to limit the bill's applicability to private sector scientists (this was not adopted) and to additionally explicitly exclude the SAPS FSL from regulation (this was adopted).

When the bill was passed that November as Act No. 27 of 2003, it included forensic science – undefined – on the list of fields to be regulated. But instead of declaring that 'This Act Binds the State', as it had in draft form, it now read 'This Act binds the State, except in so far as the State provides forensic science services'.[14] The bill as initially proposed would have defined the SAPS FSL workers as any other scientists and aligned them with the broader scientific community. Instead, it exceptionalised them as first and foremost police who used science to produce evidence for court rather than scientists whose expertise could assist the court in evaluating evidence.

The Refugees Amendment Bill emerged in 2008 in the midst of xenophobic violence in South Africa, and the problem of identification dominated its parliamentary discussion.[15] Building on South Africa's history as an innovator of the biometric state (Breckenridge, 2014), the bill originally stipulated that every applicant for asylum in South Africa must have their biometrics taken; *biometrics* was defined in the bill as 'the measurable physiological or behavioural characteristics that can be used in verifying the identity of individuals, and includes the use of photographs, fingerprints, hand measurements, signature verification, results of DNA testing, facial patterns and retinal patterns'.[16] The hope was that more data would make it possible to identify victims and perpetrators of crime when these were not South African citizens.

Two legal professional bodies, the Law Society of South Africa and the University of Cape Town Law Clinic, objected. Their concern was that collecting DNA was an unconstitutional violation of privacy for which no justification was provided; further, there was no indication of who would collect the DNA, in what manner it would be collected and where and how it would be stored.[17] To the constitutional question, a parliamentary legal advisor noted that the biometric system would only be used for refugees and not South African citizens; this raised a further issue that it would create a distinction between refugees and other people. The problem wasn't that collecting DNA was always an unconstitutional violation of privacy; it was, in fact, by then already routinely done in the course of criminal investigations. The question, then, was whether refugees would have the same rights as citizens in relation to DNA or whether they would be criminalised by being denied those rights. The refugees in question were, for the most part, Africans from elsewhere on the continent, so questions about their status in relation to criminality and citizenship contained unspoken echoes of apartheid and colonial racialised designations.

As with the Natural Scientific Professions Bill, the objections won out, to a point. Biometrics would still be collected from refugee applicants, but DNA was dropped from the definition of biometrics. This aligned refugees with citizens in terms of what biometric information could and could not be taken from them. Whereas in the Natural Scientific Professions Act, it was the SAPS forensic science laboratory that was exceptionalised, here it was DNA: since it wasn't viable to treat refugees like criminals such that their DNA could be collected when doing the same to South African citizens would unconstitutionally violate their rights, it became necessary to explicitly exclude DNA from the legal definition of *biometric*.

Debate on these bills shows how science became an open postapartheid question, sutured to that of citizenship, criminality and race.[18] In both these examples, the goals of the SAPS's use of forensic science were at odds with broader scientific consensuses about definitions of terms (*forensic science*, *biometric*) and the means of assuring scientific reliability and validity. Their success in advancing their goals, and without much parliamentary pushback, exceptionalised both the work of policing and the object of DNA as a form of evidence. Here, the potential of affirmative action and the alignment of migrants' rights with those of citizens furthered the idea that a forensic DNA database might facilitate non-racial policing.

Reimagining human rights: Criminal Law (Forensic Procedures) Amendment Bill 2008–2010

At the same time that the Refugees Amendment Bill was being drafted, debated, and finalised, an initial version of the Criminal Law (Forensic Procedures) Amendment Bill was being prepared for public comments and parliamentary discussion. The still-operative Criminal Law (Forensic Procedures) Act of 1977 had been identified as the best site for new DNA legislation for two reasons. First, this act invoked body prints and provided for blood samples, and in the absence of any mention of DNA in any existing legislation, these were taken as the best approximation of a genetic sample. Second, because of this approximation, the 1977 act had become the de facto document guiding the collection and use of DNA for forensic purposes. The 1977 act allowed for but did not require collection of fingerprints and other body prints; such samples were required to be collected by medical professionals and to be destroyed if a prosecution failed or did not go forward.[19]

This new bill, which bundled together all aspects of biometric data, had two main aims: to establish the legal grounds for searchable, digitised databases for fingerprints and DNA profiles and to shift responsibility from Health Department officials to SAPS officers for taking DNA samples. Its first iteration, drafted in 2008 and discussed throughout 2009, suggested an extreme change of course from the status quo: instead of the existing restrictive framework that subordinated crime control to the rights of the accused, it proposed an expansive framework that subordinated human rights to crime control.[20] It aimed not only to create a legal framework for a functional and cross-referenceable DNA database but also to amass the biggest database possible through expansive inclusion criteria and indefinite retention of DNA profiles.

The underlying principle at the outset, largely shaped by the perceived success of UK DNA database expansion over the previous decade, was that bigger is better: more profiles would result in higher rates of successful crime detection and prosecution. The bill proposed collecting DNA related to any and all crimes; it subjected adults and children to the same rules; and it delineated five separate DNA database indices, with different protocols for inclusion and retention. A proposed crime scene index would indefinitely store samples and any profiles derived from them. A reference index would compel samples and profiles from anyone suspected by police of involvement in any crime; samples would be destroyed after five years if no convictions took place, but profiles would forever remain on the database to search against future crime scenes. A convicted offenders index would sample all presently convicted persons to produce profiles that would also remain part of the database indefinitely. A volunteer index would be open to any adult or child, but once initial consent was granted, it could not be revoked, and these profiles would be available to search in efforts to resolve future crimes. Finally, there would be an elimination index for personnel, contractors and suppliers.

When the ad hoc National Assembly committee tasked with discussing, potentially amending and passing the draft legislation began their deliberations in January 2009, it seemed a foregone conclusion that the Criminal Law (Forensic Procedures) Amendment Bill would pass quickly and easily: the Justice, Crime Prevention and Security Cluster had in 2008 listed drafting the bill among its top priorities, thus affirming strong parliamentary support. It was framed as merely updating legislation written in 1977 that could not have included DNA as part of forensic procedures because, at that time, DNA was not part of forensic science, thus minimising the specific issues raised by DNA technologies, and it also increasingly had support among politicians who saw it as a means of aligning the SAPS with national transformation *and* international policing standards. The draft bill also addressed both the racial justice issues that had arisen earlier, without naming them as such: it addressed affirmative action by accounting for a major expansion in forensic analyst hiring that would also increase pay to aid retention of these new hires, and it even more indirectly addressed equal access to police services by emphasising that an expansive database and a large-scale national roll-out of routine collection of evidence likely to yield DNA would ensure the rights of the public to live in a society without crime. Further, not only did the bill call into existence a database that already existed, but the State Information Technology Agency (SITA) had already started to develop a computerised management system for DNA analysis that would make database expansion possible.[21] But a quick and easy passage is not what happened.

Guided by the small number of critical comments among the nearly hundred submitted during the public comment period, an internal study produced by parliamentary researchers that pointed out problems that would need to be resolved before going forward, and their own hesitations and sensibilities, doubts quickly emerged among the ad hoc committee members. They wanted clarification on which DNA samples should be retained, for how long and for what purpose. They wondered if the safeguards written into the bill making misuse of DNA samples or data a prosecutable offense would be enough to protect against potential police corruption, especially since it appeared that the intention was to have all aspects of the database and collection, analysis and storage of samples under police control. They raised concerns about the SAPS's apparent lack of readiness to implement the bill and debated whether private laboratories might also contribute and whether the SAPS FSL should be required to be accredited, given their expressed desire to do so but their lack of a clear timeline to get it done. Finally, they worried that the bill might enshrine a new form of discrimination since not everyone's DNA would be taken, but they also worried that the bill as written seemed to provide for the collection of DNA without consent from nearly everybody.

The final set of concerns, that the database might potentially be a new form of discrimination against those who were profiled or that it might, in fact, end up profiling everybody, is especially important here. At issue was that the bill authorised police to collect DNA from anyone whom they suspected might be involved in a crime. This was akin to apartheid-era mandates for warrantless searches and wide-open purviews of who might constitute a legitimate suspect, constrained only by SAPS officers' sensibilities of who was suspicious. Further, since all DNA profiles would be retained indefinitely, it would institute a permanently bifurcated population (suspects and innocents) created entirely through police prerogative, with no recourse to challenge wrongful inclusion among the suspect population.[22]

This was a moment of tension between, on one hand, sensitivities to colonial and apartheid histories in which being Black was always already suspect and, on the other hand, desires to embrace what had been presented as a neutral, scientific principle of how these databases work. As one member of the committee explained, trying to reassure another, 'We're not *criminalizing*. . . . What we would like to do is populate the databases, and the manner in which you go about populating the databases is by adding prints and samples taken from minor offences'.[23] In spite of these tensions and the unease many felt with forging ahead with a bill that was certain to face challenges for violating the constitution, the bill might have passed anyway had SAPS been able to present an implementation plan that the committee found convincing. They weren't able to do this, and the ad hoc committee felt unwilling to take ownership of a bill that they felt could not be implemented.

Their tenure ended with the end of that parliamentary term, and when the National Assembly Portfolio Committee on the Police took over the bill at the start of the next term, they decided instead to split the fingerprinting aspects of the bill from the DNA aspects because they seemed less contentious, more straightforward and easier and less expensive to implement. The committee agreed that the bill would be rewritten twice: once immediately, dealing only with fingerprinting, and again in the near future to deal with the DNA provisions. The redrafting of the DNA provisions would happen only after two forms of study took place: one, systematic consideration of potential human rights violations arising from the first version and how they might be addressed in a new version, and two, a study tour to two exemplary countries, Canada and the United Kingdom, to assess emerging international best practices. The DNA bill would be delayed, but not indefinitely: the investments already made and now planned for the future assured its eventual passage in some form, even as its initial form threatened to enshrine rather than challenge colonial and apartheid racist legacies.

International best practices, outsourcing ethics and occluding race: Passing the DNA Act, 2011–2013

In the official parliamentary report of the 2011 study tour to Canada and the UK, the main rationale given for needing to split and postpone the DNA portions of the Criminal Law (Forensic Procedures) Amendment Bill was a set of concerns that arose when the National Assembly Portfolio Committee on Police, now tasked with shepherding the bill, visited the SAPS Forensic Science Laboratory in Pretoria in early 2010. They saw

> capacity constraints particularly of personnel; unused and under-utilised equipment; problems with the procedures for the taking and safekeeping of blood samples; insufficient safeguards to protect the integrity of samples; [and] mismanagement . . . including the abuse of the tender process [corruption].[24]

Whereas the initial version of the bill required enormous trust that police would wisely use and not abuse the expansive (and expensive) powers they would have been granted and that they would competently and correctly manage samples such that links they made between crime scene and suspect profiles were reliable, that trust became untenable when faced with the realities in the lab.

At the same time, the committee was already fully convinced of the necessity of having a national forensic DNA database and fully invested in the process of legislatively calling it into being. The study tour was, therefore, an attempt to learn from those with more DNA database experience how best to approach both ethical and implementation challenges. Canada and the UK were selected as destinations because of the size and longevity of their DNA databases and because of their efforts to align desires for large databases with the necessities of remaining on the right side of human rights law (Canada because of its Bill of Rights and the UK because of its responses to the Marper ruling, which challenged its former retention policies).

In each country, the South African delegation met with police, politicians and ethicists, many of whom stressed the need for training of police, accreditation of labs and keeping the database independent from police and prosecutors. They introduced the idea that for a DNA database, bigger was not, in fact, better because overly large databases were, in practice, a source of enormous expense without the expected increase in preventing or detecting crime. They therefore recommended limiting whose profiles would be loaded and for how long they would remain on the database. Furthermore, they explained, a national DNA database was a site where existing police racial and ethnic bias tends to be further entrenched as young black males in particular were disproportionately targeted and subsequently made up a disproportionate percentage of database profiles, leading to stigma and future police harassment.[25]

The report from the study tour offered concrete recommendations for moving forward, but it also posed a fundamental question: 'South Africa needs to decide what the country wants to achieve through DNA legislation'. Long-term, ongoing costs were especially concerning, and the group who participated in the tour and authored the report felt these must be considered when deciding things like how and by whom the database would be administered and how big it would be in terms of who would be included and for how long. They recommended creating both an independent oversight body for the database itself and an ethics committee that would work to safeguard constitutional rights once the database was operational. And they concluded by directly refuting points that had been made by advocates of South Africa's first iteration of the DNA Bill: they found that the existence of the database did not act as a deterrent for re-offending, and that DNA evidence in and of itself couldn't solve crimes and should be equally emphasised with fingerprints and other investigative tools rather than prioritised. Though they still advocated going forward with legislation and building a national forensic DNA database, they warned that, compared to these first-world countries they had visited, South Africa had a smaller national budget, worse facilities and a higher crime rate, so any improvements to the criminal justice system would only materialise far in the future.

The study tour report was only part of what informed the new version of the DNA Bill that was introduced in early 2013. In 2012, the Civilian Secretariat for Police produced a policy report that was based on the study tour report; a scan of forensic DNA legislation, policy and practices in the United States, the Netherlands and Brazil, in addition to those of Canada and the UK that the study tour had discussed; other existing South African legislation that could not be contradicted, like the Child Justice Act; and interviews with a series of identified stakeholders: a South African NGO that had formed with the goal of establishing an expansive national forensic DNA database and whose founder had contributed to shaping the initial version of the bill; a university criminal procedure department; the South African Society for Human Genetics; state law advisors; and the SAPS Forensic Science divisional leadership.[26]

The 2012 policy report is decidedly more positive than the study tour report: whereas the study tour report characterised a national DNA database as inherently problematic and therefore requiring tamped-down expectations, restraint in its mandate and extensive oversight of its practices, the 2012 report concluded that a DNA database, if it accounts for constitutional rights, can not only aid in fighting crime but also enhance public confidence in the SAPS and the criminal justice system. The 2012 policy report accepted the 2011 study tour report's assertion that 'a larger database does not guarantee better chances of crime solving'. But it also asserted that, while not a silver bullet, a DNA database is scientifically sound. Here, science worked as an assurance that this was not just the right, but also the necessary path forward to align South Africa with the rest of the world.

The prospects of racial and ethnic discrimination that had been outlined in the study tour disappeared in the policy report, submerged into the language of human rights. The report listed a series of human rights, guaranteed in South Africa's constitution, that had come up in the 2009 public comments and on the study tour as at risk if left unaddressed in DNA database legislation. In each instance, it asserted how each one could be addressed. Gone was the tendency of DNA databases to reflect existing racial bias; these were instead presented as pertaining to a generalised South African population equally at risk of violation. For example, the report asserted that the right to privacy could be assured through expungement of profiles, and the use of buccal swabs instead of blood samples would mean that dignity and bodily integrity were not violated.[27] Finally, the creation of an oversight board with a mandate to ensure human rights were accounted for in the future and a mandatory five-year review period 'to address gaps, technological needs and scientific developments, and human rights transgressions' framed risk and violation as problems with technocratic solutions.

The recommendations that the database should be separate from the SAPS also disappeared. This was a matter of logistical inertia: in response to the exasperated directive from the 2009 ad hoc committee to the SAPS to demonstrate that they could implement legislation before it could be passed, increased budgets had been sought and allocated, new hiring had taken place, training had been rolled out, computer systems researched and database infrastructure plans developed. The report made it clear that just as human rights concerns could be addressed, so, too, could the concerns that had been raised about the SAPS. Capacity concerns would be met with more investment, independence concerns would be addressed by housing the database in a separate SAPS department from the work of DNA analysis, and corruption and all other current or future ethics concerns would be taken care of by the oversight board.

The debates and discussions that occurred in 2013 were increasingly constrained. Some public comments questioned giving over so much control of genetic material and information to a police service that many still did not trust, and constitutional law groups continued to push for treating arrestees like innocent citizens by, for example, having shorter retention times for their DNA profiles, even while they also acquiesced that those convicted of crimes could be denied constitutional rights. In the end, these discussions positioned nearly all who voiced opinions as aligned in a common goal of transforming policing in South Africa by investing in the SAPS's scientific capacity: passage and implementation of the Criminal Law (Forensic Procedures) Amendment Bill became inevitable, with the SAPS as the inevitable builders and custodians of the database. The idea of scientific soundness and the proposed solutions to human rights problems underwrote this inevitability.

As a new DNA bill was drafted, then amended, and finally passed, discussions focused less on principle or overall value than on finding the right balance between, on one hand, aggressively empowering police and, on the other hand, remaining within the bounds of constitutionality. For example, the final law limited the kinds of crimes for which DNA should be collected to a new list of Schedule 8 offences. It mandated that all samples except those from crime scenes would be destroyed within three months of being loaded onto the National Forensic DNA Database and it provided for expunging the profiles of arrested individuals who were not successfully prosecuted. However, except in limited circumstances like setting aside a conviction or conviction of a child, convicted offender profiles would remain on the database indefinitely.

Questions about race and justice were entirely obscured in the final legislation, so much so that race did not appear at all and was only invoked (as 'physical information') in reference to the kinds of information excluded from the definition of 'forensic DNA analysis' and 'forensic DNA profile'.[28] This exclusion was a significant development because elsewhere, legislatively allowing inference of visible traits from DNA evidence had been a primary way that forensic genetics reinscribed race and reinforced racism (Hopman and M'charek, 2020; M'charek, 2008; M'charek et al., 2020; Ossorio, 2006; Wienroth, 2018). But this exclusion cannot in and of itself address South Africa's entrenched racial inequality and the role of policing and carcerality in maintaining it.

Conclusion: DNA data worlding

Legacies of racialised oppression shaped the emergence of a South African forensic DNA database but then became harder to locate and name over time. When it was first proposed by the SAPS in 1999, the potential database was imagined as aligned with active efforts to transform policing in South Africa and even as a tool of antiracism insofar as it could diversify the ranks of the SAPS forensic science division and extend police services to Black South Africans far from urban centres. But when the request for database legislation became an actual draft of legislation in 2009, the reality of a less transformed, less trustworthy, and less ready police service prompted a more careful look at the specificity of database policies. The 2011 study tour report raised the possibility that a DNA database could extend rather than challenge legacies of racism, but when the study tour report was subsumed by the 2012 policy report, this concern was lost, replaced by an account of human rights challenges and their technocratic solutions – solutions that ultimately shaped the final version of the legislation. In this way, race became an absent presence in South Africa's national forensic DNA database (M'charek et al., 2014).

South Africa's histories of racialised policing and its national reinvention as a postapartheid democratic state beginning in the 1990s demanded an especially robust legal framework to legitimise large-scale use of forensic genetics by police and especially a national forensic DNA database. However, a robust legal framework does not necessarily mean that such a database represents a complete break with the colonial and apartheid past. On the contrary, the national problems that were prioritised in parliamentary debates remain haunted by racially exclusionary belonging, all the more so as questions about race and justice became increasingly obscured by reframed understandings of who was entitled to constitutionally protected human rights and who might best represent South Africa's citizens.

The successful passage of South Africa's national forensic DNA database legislation required the consolidation of a population whose rights could legitimately be violated. This bifurcation of South Africa's people into actual or potential violent criminals and actual or potential innocent victims is both a necessary condition and an aftereffect of amassing support for this national forensic DNA database. Earlier population bifurcations were the defining logics of colonial and apartheid regimes and likewise divided those who would have rights from those who would not; these were, respectively, citizen/subject and white/non-white. This new population bifurcation, which we might understand as citizen/criminal, necessarily builds on the colonial and apartheid bifurcations that were explicitly based on racial definition and division. Its success and, with it, the legitimacy of South African's national forensic DNA database rested on suturing the interests, fears and desires of elite actors to the project of broad-based Black empowerment: safety, security and anxiety about international legitimacy were the stakes that made such suturing possible.

Race emerges differently for those on the database and those whom the database is imagined as protecting. For the protected, race emerged in two ways: first as a question of access to high-quality policing as a means of accessing justice and then as a question of equal employment opportunities in South African forensic science for Black South Africans. But for the databased, race could emerge only as a denial.

Notes

1 South Africa initiated a criminal DNA database in 1997, two years after the UK started the first database. Lack of governing legislation elsewhere did not inhibit growth of DNA databases, and in the sections that follow, I show why it did in South Africa.
2 National Assembly Police Committee, 17 November 2021, https://pmg.org.za/committee-meeting/33845/.
3 See especially the 1950 Group Areas Act, the 1950 Population Registration Act, the 1951 Bantu Authorities Act and the 1970 Black Homelands Citizenship Act.
4 In 1998, the FSL's 38 DNA analysts handled 15,000 sexual assault cases and 10,000 child protection cases. See https://pmg.org.za/committee-meeting/4090/#_Hlk466881391.

5 For a similar legislative trajectory in the Netherlands, see M'charek, A. 2008. 'Silent Witness, Articulate Collective: DNA Evidence and the Inference of Visible Traits'. *Bioethics,* 22, 519–28.
6 National Council of Provinces Select Committee on Security and Justice, 24 February 1999, https://pmg.org.za/committee-meeting/6628/.
7 National Assembly Police Committee, 9 November 1999, https://pmg.org.za/committee-meeting/4090/#_Hlk466881391.
8 National Council of Provinces Select Committee on Security and Justice, 24 February 1999, https://pmg.org.za/committee-meeting/6628/; National Assembly Justice Committee, 15 March 1999, Justice meeting, https://pmg.org.za/committee-meeting/6467/.
9 National Assembly Police Committee, 15 February 2000, https://pmg.org.za/committee-meeting/4076/.
10 Parliamentary Joint Monitoring Committee on Children, Youth and Persons with Disabilities, 19 September 2000, https://pmg.org.za/committee-meeting/267/.
11 National Council of Provinces Committee on Security and Justice, 7 March 2001, https://pmg.org.za/committee-meeting/348/.
12 National Assembly Justice and Correctional Services Committee, 4 March 2003, https://pmg.org.za/committee-meeting/2179/.
13 My discussion of the Natural Scientific Professions Bill draws from South Africa's Parliamentary Monitoring Group archive of meetings and briefings in 2003 in which the bill was discussed. See https://pmg.org.za.
14 This act has since been amended twice, first by the Science and Technology Laws Amendment Act 16 of 2011 and then by the Science and Technology Laws Amendment Act 7 of 2014. See www.gov.za/documents/natural-scientific-professions-act-0.
15 My discussion of the Refugees Amendment Bill draws from South Africa's Parliamentary Monitoring Group archive of meetings and briefings in 2008 in which the bill was discussed. See https://pmg.org.za.
16 See the draft bill as discussed on 25 March 2008: https://pmg.org.za/committee-meeting/8989/.
17 National Assembly Home Affairs Committee, 25 March 2008, https://pmg.org.za/committee-meeting/8989/.
18 For elaboration on postapartheid suturing of nature/culture to citizenship, race and xenophobia, see Comaroff, J. and Comaroff, L. 2001, 'Naturing the Nation: Aliens, Apocalypse and the Postcolonial State'. *Journal of Southern African Studies,* 27, 627–651..
19 My discussion of the 2008 Criminal Law (Forensic Procedures) Amendment Bill draws South Africa's Parliamentary Monitoring Group archive of meetings and briefings in 2009 in which the bill was discussed. See https://pmg.org.za.
20 For a discussion of similar dynamics in the Netherlands in the preceding decade, see M'charek, 2008.
21 National Assembly Police Committee, 21 May 2008, https://pmg.org.za/committee-meeting/9147/; Justice, Crime Prevention & Security Cluster Media Briefing, 30 October 2008, https://pmg.org.za/briefing/18676/.
22 On the novel creation of 'suspect populations' through forensic DNA databases, see Cole, S. A. and Lynch, M. 2006, 'The Social and Legal Construction of Suspects.' *Annual Review of Law and Social Science,* 2, 39–60.
23 National Assembly Ad Hoc Committee on Criminal Law (Forensic Procedures) Amendment Bill, 10 February 2009, https://pmg.org.za/committee-meeting/9879/.
24 Report of the Porfolio Committee on Police Study Tour to Canada and the United Kingdom (24 June–10 July 2011), https://pmg.org.za/committee-meeting/16336/.
25 For more on how existing racial bias shapes forensic DNA databases and how this bias is sometimes minimized by advocates of such databases, see Skinner, D., 2013, '"The NDNAD Has No Ability in Itself to be Discriminatory": Ethnicity and the Governance of the UK National DNA Database'. *Sociology,* 47, 976–992.
26 National Assembly Police Committee, 13 June 2012, https://pmg.org.za/committee-meeting/14550/.

27 The understanding of buccal swabs as less violating than blood samples was crucial in expanding DNA databases elsewhere as well. See especially M'charek, A., 2008, 'Silent Witness, Articulate Collective: DNA Evidence and the Inference of Visible Traits'. *Bioethics,* 22, 519–528; Toom, V., 2012a., 'Bodies of Science and Law: Forensic DNA Profiling, Biological Bodies, and Biopower'. *Journal of Law and Society,* 39, 150–166.
28 Criminal Law (Forensic Procedures) Amendment Act No 37 of 2013.

References

Amelung, N., Granja, R. & Machado, H. 2021. *Modes of Bio-Bordering: The Hidden (Dis) integration of Europe*, Singapore: Palgrave Macmillan.

Aronson, J. D. 2012. Humanitarian DNA Identification in Post-Apartheid South Africa. *In:* Wailoo, K., Nelson, A. & Lee, C. (eds.) *Genetics and the Unsettled Past: The Collision of DNA, Race, and History*. New Brunswick, NJ: Rutgers University Press.

Beinart, W. & Dubow, S. (eds.) 1995. *Segregation and Apartheid in Twentieth-century South Africa*, London and New York: Routledge.

Breckenridge, K. 2014. *Biometric State: The Global Politics of Identification and Surveillance in South Africa, 1850 to the Present*, New York: Cambridge University Press.

Brewer, J. D. 1994. *Black and Blue: Policing in South Africa*, Oxford and New York: Clarendon Press.

Cole, S. A. & Lynch, M. 2006. The Social and Legal Construction of Suspects. *Annual Review of Law and Social Science*, 2, 39–60.

Cole, S. A. & Lynch, M. 2010. DNA Profiling Verses Fingerprint Evidence: More of the Same? *In:* Hindmarsh, R. & Prainsack, B. (eds.) *Genetic Suspects: Global Goverance of Forensic DNA Profiling and Databasing*. Cambridge: Cambridge University Press.

Comaroff, J. & Comaroff, L. 2001. Naturing the Nation: Aliens, Apocalypse and the Postcolonial State. *Journal of Southern African Studies*, 27, 627–651.

Davis, A. Y. 2003. *Are Prisons Obsolete?* New York: Seven Stories Press.

Dubow, S. 2014. *Apartheid, 1948–1994*, New York: Oxford University Press.

Gilmore, R. W. 2007. *Golden Gulag: Prisons, Surplus, Crisis, and Opposition in Globalizing California*, Berkeley: University of California Press.

Hall, S., Critcher, C., Jefferson, T., Clarke, J. & Roberts, B. 2013. *Policing the Crisis: Mugging, the State, and Law and Order*, Houndmills, Basingstoke anf Hampshire: Palgrave Macmillan.

Hindmarsh, R. & Prainsack, B. (eds.) 2010. *Genetic Suspects: Global Governance of Forensic DNA Profiling and Databasing*, Cambridge and New York: Cambridge University Press.

Hopman, R. & M'charek, A. 2020. Facing the Unknown Suspect: Forensic DNA Phenotyping and the Oscillation Between the Individual and the Collective. *BioSocieties*, 15, 438–462.

Hornberger, J. 2011. *Policing and Human Rights: The Meaning of Violence and Justice in the Everyday Policing of Johannesburg*, New York: Routledge.

Krimsky, S. & Simoncelli, T. 2011. *Genetic Justice: DNA Data Banks, Criminal Investigations, and Civil Liberties*, New York: Columbia University Press.

Lynch, M., Cole, S. A., Mcnally, R. & Jordan, K. 2008. *Truth Machine: The Contentious History of DNA Fingerprinting*, Chicago: University of Chicago Press.

M'charek, A. 2008. Silent Witness, Articulate Collective: DNA Evidence and the Inference of Visible Traits. *Bioethics*, 22, 519–528.

M'charek, A., Schramm, K. & Skinner, D. 2014. Technologies of Belonging: The Absent Presence of Race in Europe. *Science, Technology & Human Values*, 39, 459–467.

M'charek, A., Toom, V. & Jong, L. 2020. The Trouble with Race in Forensic Identification. *Science, Technology, & Human Values*, 45, 804–828.

Machado, H. & Granja, R. 2020. *Forensic Genetics in the Governance of Crime*, Singapore: Palgrave Macmillan.

Machado, H., Granja, R. & Amelung, N. 2019. Constructing Suspicion Through Forensic DNA Databases in the EU: The Views of the Prüm Professionals. *The British Journal of Criminology*, 60, 141–159.

Mamdani, M. 1996. *Citizen and Subject: Contemporary Africa and the Legacy of Late Colonialism*, Princeton, NJ: Princeton University Press.

Ossorio, P. N. 2006. About Face: Forensic Genetic Testing for Race and Visible Traits. *Journal of Law, Medicine & Ethics*, 34, 277–292.

Posel, D. 1991. *The Making of Apartheid 1948–1961*, Oxford: Clarendon Press.

Skinner, D. 2013. 'The NDNAD Has No Ability in Itself to be Discriminatory': Ethnicity and the Governance of the UK National DNA Database. *Sociology*, 47, 976–992.

Skinner, D. & Wienroth, M. 2019. Was This an Ending? The Destruction of Samples and Deletion of Records from the UK Police National DNA Database. *BJHS Themes*, 4, 99–121.

Toom, V. 2012a. Bodies of Science and Law: Forensic DNA Profiling, Biological Bodies, and Biopower. *Journal of Law and Society*, 39, 150–166.

Toom, V. 2012b. Forensic DNA Databases in England and the Netherlands: Governance, Structure and Performance Compared. *New Genetics and Society*, 31, 311–322.

Vailly, J. & Bouagga, Y. 2019. Opposition to the Forensic use of DNA in France: The Jurisdiction and Veridiction Effects. *BioSocieties*, 15, 394–419.

Wienroth, M. 2018. Governing Anticipatory Technology Practices: Forensic DNA Phenotyping and the Forensic Genetics Community in Europe. *New Genetics and Society*, 37, 137–152.

PART II

New and emerging innovations and applications

6

FROM PROMISE TO PRACTICE

Anticipatory work and the adoption of massive parallel sequencing in forensics

Roos Hopman, Irene van Oorschot and Amade M'charek

A victory – for whom?

17 January 2019 marks an important day. It is an important day for our suspect, P., who is found guilty of raping a young woman and sentenced to three years in prison. This decision in the Brettenpad case overturned a 2017 ruling that had acquitted P. due to insufficient evidence. The appeal decision therefore came as a win for the prosecution. Moreover, a novel forensic genetic technology was crucial to bringing about this appellate decision, marking this the first forensic case in which the application of massive parallel sequencing[1] (MPS) technology led to a conviction. With the application of MPS, the short tandem repeat (STR) profiles constituting the central pieces of evidence in this case could be attributed to the victim and suspect P., allowing the verdict to shift from '*pointing towards* involvement of the suspect'[2] after the first hearing towards '*proven that* the suspect committed the criminal fact'[3] after the second.

Although MPS was announced as 'The Next Big Thing in Genetic Medicine' ten years ago and has been applied routinely as a diagnostic tool in medical genetics (Tucker et al., 2009), it is a relative novelty to the field of forensic genetics (De Knijff, 2019; see also Wienroth et al., 2014, on the 'fourth wave' of innovations in forensic genetics). Even though attention to the technology in the forensic genetic field has increased over the past few years, up until the Brettenpad case it had not yet led to any breakthroughs in criminal cases. Indeed, its prior mobilisation in a US case turned out to be rather unsuccessful, as the judge in the case ruled that MPS was a non-admissible technique due to a lack of forensic validation.[4] The lack of MPS applications in forensics is striking as its ability to reveal DNA variation in great detail would be uniquely suited for forensic casework (De Knijff, 2019) where DNA material tends to be scarce. In rape cases in particular,[5] DNA materials tend to come in complex mixtures of two or more donors (cf. Lander,

DOI: 10.4324/9780429322358-9

1989; Thompson, 1993), bolstering MPS's forensic appeal. In contrast to capillary electrophoresis-based STR analysis, the current gold standard in forensics, MPS not only provides insight into DNA fragment lengths but also reveals the underlying base pair variation (De Knijff, 2019, p. 176). Therefore, in criminal cases involving complex DNA mixtures such as the one presented in this chapter, MPS-based analyses allow for the different donors to be distinguished. As the eventual conviction of P. demonstrates, this can prove essential to establishing a suspect's guilt.[6]

In this chapter, we ask how legal and forensic actors created the necessary space for MPS to go from promise to practice. What were the actors involved in this translation? How did they manage to bring MPS to bear on this specific case? We bring into view the decisive stage in the implementation of a new forensic technology, demonstrating the translation of a *promissory* technology into an *applied* one. Through the Brettenpad case, we show that this translation depends on and is rendered possible by anticipatory work practices. These practices, we show, are distributed over a variety of actors and settings. They are also crucial to the configuration of what Wienroth (2018), following Ulucanlar et al. (2013, p. 98), calls an 'adoption space':

> [A] spatial and temporal space . . . populated by human and non-human actors . . . where attitudes, practices, interactions and events, together with the technology's material features, shape technology perceptions in ways that are instrumental in decisions about its use.
>
> *(Ulucanlar et al., 2013 cited in Wienroth, 2018, p. 138)*

In our analysis of the more or less 'invisible' (Star and Strauss, 1999) anticipatory practices involved in making MPS 'ripe for legal use', we emphasise these multiple actors and technologies, showing also how serendipity, receptiveness among the judiciary and a forensic enterprising spirit were essential to translating this promissory technology into an applied one. In doing so, we build on Wienroth's (2018) understanding of 'anticipatory work', as practices involving (discursive) promise making, but also epistemic and operational dimensions. We draw from formal documents, interviews with a forensic expert and field notes taken in the forensic laboratory and during court hearings to emphasise the advocating, contestation, experimentation, patience but also serendipity essential for MPS to have culminated in this conviction. In doing so, we add further complexity to Wienroth's understanding of anticipatory practices, stressing not only the deliberate, goal-oriented 'work' done by key actors pushing for the implementation of novel technologies but also less visible tinkering practices. More experimentational, these tinkering practices are 'surprising and adaptable', demanding an 'attuned attentiveness' (Mol et al., 2010, p. 15) from different actors but, as we demonstrate, from lab technicians in particular.

In the following, we first situate MPS vis-à-vis established forensic identification techniques, after which we turn to a discussion of the conceptual tools we bring to this crucial technolegal event. We then discuss our methodological engagement with this case in more detail. Following this, we take the reader from the court to the lab and back, demonstrating the ways various actors became instrumental in the creation of an 'adoption space', highlighting both partially invisible tinkering practices (Star and Strauss, 1999) and more overt forms of anticipatory work and entrepreneurship. In the conclusion, we reflect on our findings, commenting in particular on the technolegal worlds rendered possible through this technology.

Forensic promises and massive parallel sequencing

Forensics and MPS

Forensic identification's prime goal, as noted by the geneticists Jobling and Gill (2004, p. 739), is 'to identify with as much certainty as possible the origin of a biological sample'. In other words, it seeks to answer the question, *Who* can a sample be attributed to? Since roughly 1995, the most commonly used way to identify the donor(s) of a sample has been through capillary electrophoresis–based short tandem repeat (STR) analysis, also known as DNA profiling or DNA typing. As the term 'short tandem repeats' suggests, STRs are short sequences in the DNA that repeat themselves in the same pattern but in differing numbers. One individual might have five repeats of a specific sequence at a certain location in the genome, whereas another individual will have seven. Because of this variability, STR markers have a discriminatory potential and are ideally suited for forensic application (M'charek, 2000). By typing a set of around twenty loci or markers (depending on the kit that is used) and documenting how many repeats a person has for each of them, an individually 'unique genetic code', or DNA profile, can be generated (Roewer, 2013). These DNA profiles are considered individualising to such an extent that they are said to provide a 'DNA fingerprint' of a person (Jeffreys et al., 1985).

With capillary electrophoresis analysis of STRs, the variation in lengths of fragments of DNA can thus be revealed. Yet when dealing with mixed DNA samples, which is common in forensic cases, it may be impossible to distinguish all donors. First, there is the issue of homoplasy, referring to DNA fragments that are of the same size but differ in their composition. Here, fragment length does not provide enough information and cannot help distinguish different donors. In a mixed sample to which several donors contributed, homoplasy hence complicates the interpretation of the output as some contributors might carry different alleles that have the same size. Second, capillary electrophoresis is prone to errors. The most frequent of these errors, 'slippages' that occur during the reaction, cause 'stutter' alleles to appear in the output (De Knijff, 2019, p. 176). These stutter alleles visually present themselves as minor peaks, often right before the proper allele. Stutter alleles are especially problematic in mixed forensic samples, where they become indistinguishable from alleles stemming from minor contributors (Budowle et al., 2009). This is an issue in rape cases in particular, as here the victim's DNA is often present in much larger quantities than that of the perpetrator.

On both counts – the problem of homoplasy and that of stutter errors – MPS-based analysis of STRs can make a difference. In addition to revealing the lengths of

DNA fragments, MPS makes insightful the sequence variation of these fragments. MPS does not type STRs in the sense that it produces output reporting whether a donor had variant A of a marker or variant B, but instead sequences the alleles and reveals their nucleotide composition. Through this, substitutions occurring at a single base (single nucleotide polymorphisms or SNPs) in the repeats can, for example, be detected. In the case of homoplasy, this allows for the distinguishing of alleles of the same length through studying their sequence composition. Stutter alleles, while not completely surmountable, furthermore become less problematic as comparisons with population data enable lab technicians to distinguish 'noise' from genuine alleles to a greater extent. (We will elaborate on this later in the chapter.) Consequently, MPS affords greater precision to the DNA profile and subsequently adds increased discriminatory power to STR markers that are already in use in forensic casework. In contrast, then, to the emerging technology of forensic DNA phenotyping, which, in its grasp for individually specific facial composites of unknown suspects, produces racialised collectives (Hopman and M'charek, 2020; M'charek, 2020; see also Cole and Lynch, 2006; Ossorio, 2006; Skinner, 2020; Wienroth et al., 2014), MPS is a technology that contributes to the individualisation of forensic DNA analysis (cf. M'charek, 2000, 2008). Building on the historical promise of DNA as a 'truth machine' (Lynch et al., 2008; Toom, 2011), MPS promises further specificity and individuality.

Anticipatory work as tinkering work

If the preceding paragraphs offer an overview of the forensic promise of MPS, we must also bear in mind that these promises require a host of what Wienroth (2018), in discussing forensic DNA phenotyping, calls 'anticipatory work'. Crucially, this work is based not only on discursive action (e.g., the making of promises, the communication of insights) but also on operational and epistemic practices. Operationally, the technology in question must be ironed out before or through testing, or, rather, its specific limitations and error rates must be rendered known in advance. This work is both technical and performative, in that developers aim to 'perform legitimacy and social safety' (2018, p. 144). As we emphasise in our analysis, to a large extent this work is experimental and playful, requiring technicians to be attentive to the specifics of a particular case and lab machinery to adjust their analyses accordingly. We therefore show that, similar to care practices, 'rather than requiring impartial judgements and firm decisions', anticipatory practices 'demand attuned attentiveness and adaptive tinkering' (Mol et al., 2010, p. 15). Such tinkering moments are precisely the space in which (the legitimacy of) potential future uses and possibilities are teased out and so are crucial to apprehending the way promissory technologies are made into a reality.

Epistemically, too, work has to be done: in order for the technology to be used in practice, various actors must shape the epistemological and ontological scene onto which it may appear (2018, p. 144). In other words, technologies must manage to resonate with existing problematics (in this case, the difficulty with interpreting complex DNA mixtures) while simultaneously teasing out a space of novelty within which existing problems may be cast in a new light or approached from a different direction. Taken together, Wienroth suggests, these dimensions of anticipatory work may create an 'adoption space' (Ulucanlar et al., 2013) within which a specific technology may then be 'settled in' to become part of forensic infrastructures of identification. We do not aim to give an exhaustive analysis of this adoption space but to focus on key epistemic anticipatory work in the shape of tinkering and to demonstrate the impact of that work for part of the operational dimension of introducing a novel technology.

While our case shows that different actors are, indeed, anticipating and working towards the adoption of MPS in forensic practice, it brings to the fore two additional dimensions, adding further complexity to Wienroth's conceptualisation of anticipatory work. One, we demonstrate the fundamental openness and excessiveness of the future itself. In so doing, we zoom in on the futures evoked and mobilised in forensic practices, hence thematising the role of temporality in such work. We may understand this work and the promissory nature of MPS better through the writing of Mike Fortun. Quoting Hans-Jörg Rheinberger (1997), himself quoting the geneticist Francois Jacob on the history of experimental science and the role of epistemic things, Fortun states that the sciences tend to be '"a machine for inventing the future" – because, paradoxically, they've already downloaded part of that future into themselves, via promising' (Fortun, 2005, p. 165). While partially folded into the machinery of science, this future cannot be reduced to a present or a past but is produced in excess, echoing the saying that 'the future is full of possibilities'. Secondly, then, our case demonstrates the concrete spaces and practices within which multiple future possibilities are tested out: in other words, where the excessiveness of the possible[7] is worked out and through. In our case, tinkering with and testing the apparatus in the technicians' practices come to the fore as moments in which such possibilities are teased out and so are important moments to take into account when explaining the translation of MPS into forensic practice. Throughout our analysis, then, we emphasise the distributed and partially invisible character of anticipatory work while, at the same time, remaining attentive to individual forms of entrepreneurship as well as non-theorisable forms of serendipity and timing.

MPS and the Brettenpad case

MPS became legally decisive for the first time in the so-called Brettenpad case, making this a forensic case to take note of. In the early evening of Mother's Day, 10 May 2015, a 27-year-old woman was cycling home along the Brettenpad, a cycling path running through a nature park in the vicinity of Amsterdam. Suddenly, a man started cycling next to her. He dragged her off her bike and raped her. After the rape, the victim sampled her body with cotton buds and assembled the clothes she had worn in a plastic bag. She handed these items, plus her cell phone that the perpetrator had touched, over to the police. The items were searched for DNA traces, and the collected DNA was sent to a commercial forensic laboratory (from here onwards, lab A) in the Netherlands. Between October 2015 and April 2016, lab A performed capillary electrophoresis–based STR analyses on the traces in order to generate STR profiles. The STR profiles, all mixtures, were compared to profiles in the Dutch national forensic DNA database in search of matches. This is when suspect P. first came into view: one of the mixtures generated a partial match[8] with his DNA profile. On 12 May 2016, the suspect's legal counsel was contacted and informed of the suspicions against his client. In response, the suspect reported himself to the police on 17 May. Because of the complexity of the profiles, the defence requested counter expertise to be performed on the traces by another Dutch forensic lab (lab B), the results of which are dated 17 January 2017.

When presented with the results of the STR analyses done by lab A, however, the head of lab B was not convinced that DNA profiling would be the most insightful technique for this case. Because the generated profiles were mixtures compiled of contributions by several donors, distinguishing which markers belonged to whom was complicated. Importantly, this expert therefore proposed to conduct MPS instead, as he was certain that this technology could make 'crystal clear' whether the suspect had contributed to the mixture or not, as he emphasised during a conversation at his lab in April 2017. Crucially, he had informed the defence of his appraisal of the mixtures and pointed out the discriminatory power of MPS, making sure to also address the potential negative consequences it could have for defendant P. MPS, he suggested, could lead to clear exclusion of the defendant, but it might also consolidate the match between the traces and suspect P.'s profile. He felt it was his responsibility as a DNA expert to make sure the defence was aware of the potential consequences of the analysis: 'Lawyers can't know everything'. The defence decided against MPS analysis of the traces and instead requested that the counter expertise be conducted with the same techniques as performed by lab A.

Besides the DNA profiles and the resulting partial match in the DNA database, clues in the case were limited to three other pieces of circumstantial evidence. First, the victim had described the perpetrator as a 'white male' of 'normal posture' with a 'dark brown beard', possibly of a 'Middle Eastern background'. The court, however, ruled that even though the suspect fit the description, many other men would as well. Since no distinctive characteristics of the perpetrator were mentioned, the witness description was deemed too generic to identify the defendant and consequently insufficient – a problem to which such witness descriptions are particularly vulnerable in legal practice (van Oorschot, 2020). Second, the alibi the suspect provided was deemed 'not credible' by the court and was contradicted by his mother and a friend with whom he had spent time on the day of the crime. Third, even though the suspect proclaimed never to use a bicycle, his friend testified to having seen him cycling. Furthermore, this friend noted that the suspect's bicycle fit the description the victim had given of the perpetrator's bike, which was a 'woman's bike'.

While these three clues pointed to a possible 'involvement of the suspect', they remained insufficient to convict the individual. The Dutch criminal law system does not operate with the (US) 'beyond a reasonable doubt' criterium but rather insists that the defendants' guilt must be *wettig en overtuigend* [lawfully and convincingly] proven. While these three circumstantial forms of evidence may have generated some 'conviction' among the three judges of the *meervoudige kamer* [judicial panel], they were not sufficient to lawfully establish his guilt. Moreover, expert B suggested that the fact that the defence had decided not to opt for MPS may have raised suspicion among the three panel judges as well. MPS, however, was not yet ordered as the court reasoned that this decision was to be taken by the public prosecution, not by the court itself.

Configuring an adoption space, putting MPS to the test

While the panel of judges decided to acquit the suspect on 30 March 2017, the court session nevertheless played a crucial role in configuring the 'adoption space' for MPS. After all, interpreting the initial STR profiles was not straightforward. The prosecutor and the chair were unfamiliar with the DNA terminology used in the reports and struggled to make sense of the findings. Most significantly, the experts agreed that the largest contribution in the mixtures came from the victim herself, with minor contributions from one or more other donors. Yet on how many donors contributed and which markers belonged to whom, the experts did not agree. As the expert heading lab B would tell the first author of this chapter, hereafter Roos, later: '[Y]ou simply cannot make a decision in my opinion; you cannot objectively choose between the two scenarios'. As we will see, this disagreement among experts, while puzzling, would create the necessary epistemological reflections for MPS to be put forward as a solution to the problems adhering to conventional STR profiles. Here, our more detailed observations in court testify to the way in which the expert witness, the head of lab B, managed to elicit interest in the technology as an alternative way of generating information about the suspect's involvement.

A tale of two scenarios: navigating expertise, raising possibilities

The hearing, which started on a Thursday morning and lasted well into the afternoon, revolved around the expertise of the two DNA experts introduced as 'a forensic DNA expert' (expert A), and a 'professor in population and evolutionary genetics' (expert B). In the Dutch criminal justice system, it not especially common for experts to appear in court: forensic knowledge travels via the official dossier – a crucial actor in the Dutch context – so that the behind-the-scenes work (interactional, technical) tends to be black boxed and rendered invisible (van Oorschot, 2014, 2018, 2021).[9] However, this complex case involving counter expertise called for their physical presence in court (see also M'charek et al., 2013). The first question the chair posed to the experts concerned the methods they had used to generate their STR profiles. She asked them to explain what they had done and to elaborate on the differences between their approaches. She also requested them 'to keep it basic', emphasising that 'we are not advanced students in genetics'. The experts subsequently explained their approaches.

Expert A elaborated that in his lab, STR profiles were generated based on 16 loci.[10] To decide to whom the alleles that were found in the profiles belonged, they fed these mixed profiles into a software program. The program was designed to calculate the likelihoods of there being a particular number of donors and, in this case, the likelihood that the suspect was one of them. The calculations were based on a priori defined scenarios that were fed into the software before the analysis. What it produced were thus the likelihood ratios for a set of pre-determined scenarios, each differently explaining the composition of the donors in the mixture. Expert A explained: 'The program gives the most likely profiles for each sample'. Based on samples taken from the rear of the victim's underwear, they concluded that the scenario that the DNA mixture contained contributions of the victim, the suspect and an unknown person was 'extremely more likely' than the scenario that contributions came from the victim and two unknown persons. The conclusion from their analyses was therefore that suspect P. could not be excluded from the mixture.

Technicians in lab B looked at 22 loci for each of the samples. They did not make use of a software program to calculate likelihoods; during the hearing, expert B stressed he was no proponent of this method as it requires 'an unnecessary extra translation' and is based on 'a priori assumptions'. Instead, they produced what they called a 'consensus profile' by typing the traces four times with two different kits. Markers that were present in each of the typing runs were included in this consensus profile. To determine whether the suspect could be included in the mixture, they compared it to the suspect's profile. From the comparison, expert B was unable to conclude whether to include or exclude suspect P. He deemed the samples too complex and the quality of the DNA material too low to draw definite conclusions.

The chair took the time to make sense of the differences between the methods used in the two labs and the significance they could have for the results. She asked many questions: for example, what it means to consolidate a 'match' and if 22 loci necessarily yield more information than 16. She additionally inquired if it made any difference that in lab B, traces were typed four times, asking what consequences this could have for the resulting consensus profile. Expert B explained that it could, indeed, make a difference in 'complex cases such as this one' as 'some kits will make the perpetrator surface and others won't'. In her questioning, the chair actively sought to understand the terminology used by the experts, their respective methods and the resulting findings. In a context in which forensic results, by virtue of being delivered in the shape of a report written up by an accredited institution, tend to be 'black boxed', the chair's extensive questioning is remarkable. After having thoroughly covered the methods, the discussion moved on to the reported results.

This part of the discussion mainly revolved around the number of donors who contributed to the sample and whether the suspect could be one of them. Consider the following interaction between the chair and the two DNA experts:

EXPERT B: We were unable to include or exclude the suspect because there was insufficient DNA found in the mixture.

CHAIR: Exclusion sounds very absolute. Does that mean that it is impossible for him to have played a role in this case?

EXPERT B: When I use the word 'exclusion', it means that we were unable to retrieve him from the profile.

CHAIR: You are *not* saying that it is *impossible* for this material to belong to this person?

EXPERT B: We did not find clues for that. That is all I can say.

EXPERT A: This is the reason I never use the word 'exclude'. It is confusing. We reach another verdict with our analysis.

EXPERT B: My colleague is more cautious.

Here, we see the discussion of the results pivoting on the notion of exclusion. Throughout the discussions on the profiles and the differing opinions of the DNA experts, the complexity of the profiles became evident – but it also generated questions about what it means to 'exclude' a scenario based on DNA evidence alone or in relation to the 'context of the crime'. Anticipating these difficulties, expert B had made a note on the potential of MPS to provide more clarity in his written report to the court. While this technology was not in use, the chair moved the discussion on to MPS. Here, again, the chair was thorough in her questioning.

CHAIR: Could MPS add something we don't know yet? What does this method do?

EXPERT B: You can see things differently with MPS. It is a completely different technique to make the same traits visible. With that, you can get a better understanding of who the donors are. With the use of MPS, the sensitivity can be much higher, so it could give clearer answers.

CHAIR: So it is about making the same traits clearer?

EXPERT B: Yes.

CHAIR: But it does not generate more information?

EXPERT B: It does. With MPS you take another perspective. You don't look at the variations in length but at the composition of the sequence. By doing this, the rarity value per trait increases. You can distinguish donors more easily. So what I am saying has two dimensions. With MPS, it becomes easier to interpret what we already know, and then we can determine whether the visible peaks are stutter peaks belonging to the victim or peaks belonging to the suspect. There are specific cases in which this can be valuable. It can both benefit and work against the suspect. This is a consideration you have to make.

CHAIR: Could MPS completely change the results you have found?

EXPERT B: That depends on the amount of markers that were included in the match. But I expect that the match will be consolidated.

In this exchange, it becomes clear how expert B actively advocated for MPS as a useful technique through emphasising its 'anticipated benefits' and added value (Wienroth, 2018, p. 5) for this case: namely, adding more precision and detail to the profiles. In his efforts, he was helped by the inquisitiveness of the chair and prosecutor. After the chair questioned expert B on MPS, the prosecutor requested him to re-explain STRs and to address the meaning of 'likelihood ratios', and finally, she tried to make sense of the lack of consensus between the experts by asking them to comment on each other's methods. During a break in the late afternoon of the hearing, expert B told Roos he was surprised by the thoroughness of the chair's and prosecutor's questioning, noting that 'usually they only ask one or two trivial questions'. Here we see how anticipatory work is also work of sensitising the court to different 'perspectives', to 'seeing things differently' using MPS, hence doing the epistemic work imperative for a technology to become framed as necessary, even desirable. Not just in terms of technical precision but also because it would feed into the delivery of justice in complicated rape cases such as these, so it inherently promises increased social safety.

The court took the disagreement between the experts, in particular expert B's hesitation to draw conclusions based on the DNA materials, very seriously. They considered it 'irresponsible' to attribute 'decisive meaning' to the likelihood-based findings of expert A. The court therefore acquitted suspect P.: it ruled that the results pointed towards his involvement in the rape but were insufficiently convincing to identify him as the perpetrator.

Importantly, the court took up expert B's mentioning of MPS in its written verdict. It mentions that MPS could increase the *bewijskracht* [evidentiary value] of the profiles:

> The court realises that, in case it were to call for further forensic research using 'massive parallel sequencing' (MPS), a technology in which not only the length but also the composition of individual building blocks of the DNA are taken into account, the evidentiary value of the tested DNA profiles may increase, as the discriminatory value of the DNA profile may be enhanced. [Lab B] is the first accredited lab in the world to apply this technology.[11]

While highlighting the promise of MPS, then, the court did not want to go 'as far [as] to order for the gathering of new evidence or to increase the evidentiary power of existing materials' as this would be 'taking the position of the prosecutor', hence nudging the prosecutor to file for appeal. Here, the boundary between the prosecution and the court is asserted, yet it is done in such a way as to signal the court's receptiveness to the mobilisation of MPS.

Looking back with expert B

Three months after the hearing, in June 2017, Roos attended a conference on forensic genetics in Dubrovnik, Croatia, where she had the opportunity to have a few informal conversations with expert B. By this time, the public prosecutor had indeed filed for an appeal and ordered the application of MPS to be performed by lab B. During one of their conversations, they spoke about the case and its complexities while also addressing the added value MPS would have for it. Expert B explained that he was working on more MPS cases but that this one was especially interesting. 'Everyone', he said, 'is of course assuming that MPS will give us so many clues that he [suspect P.] will fit the profile, that he will still be prosecuted'. Here, it was not only the three pieces of circumstantial evidence that played a role in generating strong suspicion but also the defence party's refusal to allow lab B to apply MPS. A shame, really, as expert B saw this case as 'the prototype case' in which, based on capillary electrophoresis, 'You simply cannot make a decision in my opinion; you cannot objectively choose'.

MPS, he stressed, would most likely provide the means to make this decision. Even though he believed strongly in the power of the technology to make a difference in this case, calling it a 'no-brainer' to apply MPS to the mixtures, he emphasised that waiting for the right moment before applying a forensic technology for the first time is crucial: 'You have to wait until the time is ripe'. The translation of a promissory technology into an applied one is thus importantly also about being patient and knowing the right time and case to put it into practice. This geneticist mentioned a complicated twin case in the US that had been based on MPS, which had been dismissed by the judge. He therefore stressed caution when introducing a technology because when not done properly, it will take a lot of work to get it introduced at all. By working on this case, he was hoping to give the forensic world the right example:

> That's why companies that deliver products are really keeping an eye on me, and why all my colleagues know what I am doing. Most probably I will be the first in the world to apply MPS under accreditation in an X number of cases. And be able to show it.

This quote and the importance of 'getting it right' on the first application demonstrate the political economy of forensic genetics as the stakes of scientific, forensic and commercial actors intersect (Wienroth, 2020).

Despite the expectation that in the Brettenpad case MPS would consolidate the match between suspect P. and the DNA mixture, expert B emphasised that he needed to keep 'all his options open'. The ability to make an 'objective' decision was important to him. He emphasised that, for him, it was crucial to 'have integrity' when doing forensic casework and to perform the analyses without any presumptions in mind.

With that, we move on to the laboratory, where various lab technicians were busy ironing out the technology for use. Turning towards this operational dimension (see Wienroth, 2018, p. 145), we show how these largely invisible work practices (Star and Strauss, 1999) were instrumental in working out its affordances and limitations.

Tinkering in lab B: The game of artefacts

In the early spring of 2018, Roos conducted two months of field work in the laboratory of expert B. Here, she was able to witness some of the very first runs of the MiSeq (a particular DNA sequencer[12] used to conduct MPS) on forensic traces[13] and to be present for the initial analyses and interpretations of data produced with MPS. She accompanied one of the technicians who prepared DNA materials to run them on this sequencer. With this technician, she was able to observe the technical process that preceded the sequencing of the DNA material with the MiSeq in its entirety, a complicated and time-consuming series of procedures that took three days in total to complete. The technician conducted this cluster of tests, consisting of the copying, labelling, cleaning, quantifying and pooling of DNA, by herself. At any point during the preparation steps, minor things could go wrong, which would subsequently tamper with the sequencing output. This technician therefore emphasised that when she was still an intern at this lab, she was not so eager to perform MPS: 'There is so much that can go wrong. And you can only tell at the very end, so then you have to completely redo everything. On top of that, every run costs a few thousand euros'.

After the preparatory steps were completed, the samples could be mounted onto the sequencer. The MiSeq then took an additional 56 hours to sequence the samples. When the sequencing was successful, the results were passed over to two other lab technicians. For their part, they went carefully through the output using 'FDS tools', software designed to help interpret and analyse the sequencing results. For each sequenced marker, the software reported which variants were found (in this case, 24 markers were sequenced for an autosomal profile and 42 for Y SNPs)[14] and, in addition, displayed the sequence compositions of each STR. Most importantly, the analysis of these results then revolved around 'taking out all the artefacts'. The technicians therefore had to find ways to determine whether the markers displayed by the software were 'genuine' alleles or 'noise'. Were they looking at an allele belonging to a human donor? Or was it simply stutter? Roos had the opportunity to sit with them as they discussed each allele.

Deciding whether a marker was genuine or artefact was complicated, in particular because the DNA in the samples was degraded and of low concentration. Signals from minor contributors were often indistinguishable from stutter. While discussing a particularly ambiguous marker, one of the technicians referred to the analysis as an '*artefacten spel*' ['game of artefacts']. In order to make the distinguishing of artefacts from genuine alleles easier, FDS tools gave 'hints' based on a reference database. In this case, 700 reference profiles had been fed into the software, based on which it predicted which alleles were real (because they occurred in the reference data) and artefact (because they occurred in a locus where noise was common in the reference data). Furthermore, when the technicians were unsure about an allele, they made manual comparisons with their own additional database of 2,000 individuals. If an allele occurred in this database, they deemed it more likely they were dealing with a 'real' allele. Nonetheless, if an allele did not show up in the database, there was no guarantee that it could not be genuine. The technicians remained uncertain:

> It is possible you have found a new allele, but you can also find artefacts. Which one of the two options it is you don't know. That's why we do a second PCR on some markers to double check. Nothing is certain in this analysis. All we get are hints from the software and the databases, and based on that we have to make decisions.

Being among the forerunners sequencing forensic DNA traces using MPS, these technicians had to develop a method for analysing the results completely from scratch. There was no existing nomenclature for reporting the found variants and no previous forensic casework to rely on. This was also the reason they worked together: one of the technicians had already taken out the artefacts for each marker beforehand but went through the 'raw' output again with their colleague to re-analyse the alleles, allowing them to discuss ambiguous markers. Analysing the samples in this manner was slow work. It initially took the technicians hours to go through a few markers. For each marker, the first technician would report which alleles she had evaluated as artefacts, after which the second technician would look at the found alleles and re-evaluated the assessment. The examination of every single marker therefore culminated in a discussion.

T1: It looks like junk. But I think there might be something behind it.
T2: I was doubting a lot. I think [the signal] is very high for junk.
T1: Leave it in. What are these extracts actually?
T2: Sample from the shirt or the underwear or something.
T1: Old material?
T2: Some years ago, yes. I'll look it up. . . . This is from the vest, and it is 1996 we're looking at.
T1: Is the concentration low?
T2: We didn't perform a measurement on the samples.
T1: You can clearly tell these samples are inferior. Lower input, worse material, so higher artefacts. We have seen this one [allele] four times now. I don't think it is real.
T2: But just now you said that we should leave it in.
T1: Yes, the first two times we encountered it. That eight is high, but still I think it was an artefact. Yet if I would take it out, I wouldn't be able to defend that choice.

Aiming to make MPS a workable, operational technology, in the lab, the technicians tried to navigate and 'manage' the uncertainty of the analysis (Kruse, 2016, p. 70). They did so by building on hints provided by the software and databases, but as becomes clear from the excerpt here, they also needed to develop a sensitivity to the particular PCR that was used prior to the sequencing. After comparing results from several runs, they found that at particular markers, this PCR was prone to producing stutter. They subsequently took this into account during their analysis. Furthermore, the concentration of the samples, where the samples stemmed from and how old they were also became relevant to distinguishing artefacts from true alleles. To draw a parallel with a classical review of DNA typing, making decisions therefore required that these lab technicians have not just an expert knowledge but also a highly situated and intimate knowledge of their instruments and contexts: indeed, 'a sophisticated knowledge of both the procedures used to create the prints and factors that could cause variability in the results' (Thompson, 1993, p. 42). Being the first lab to apply MPS to forensic traces and having to work from scratch, these technicians thus had to tinker with the technology and the results it produced. It required them to 'try again, try something a bit different, be attentive' (Mol et al., 2010, p. 14). Tinkering practices as such prove crucial to understanding the ways in which technicians speculate on future possibilities and uses and how the promissory character of MPS is being worked out and through backstage.

Indeed, it is through these initial analyses of forensic case material that the foundation was provided for later work, such as that on the Brettenpad case. For the Brettenpad case, eight DNA traces were analyzed at this lab for 'reliable' and 'reproducible' genotypes. This meant that, as described earlier, artefacts were distinguished from genuine alleles for each of the markers found in the mixed samples. Analysis of three of the traces resulted in reliable MPS profiles consisting of 22 markers each. Subsequently, these were compared to the MPS-based DNA profiles of the victim, the victim's partner and the suspect. Lab B performed these analyses and handed in the resulting report on December 7, 2018. These results then became decisive in the early 2019 verdict.

Back into court: The appeal

A court hearing revolving around the new results took place on December 20, 2018. As reported during the hearing, MPS analysis of the traces resulted in three mixed profiles. To explain each of these mixtures, lab B's report proposed two alternative hypotheses:

1. The results obtained for the trace can be explained by mixing cell material of the victim, the suspect and 0–1 unknown persons who are not related to the victim or the suspect.
2. The results obtained for the trace can be explained by mixing cell material of the victim and 1–2 unknown persons who are not related to the victim.

As was repeated in court, the report then went on to stress that the results were 'ten billion times more likely under hypothesis A'. Taking these results into account, 'especially considering the rarity of the MPS based profiles', the court ruled that the trace material found at the crime scene must partially have stemmed from the suspect. From that point onwards, the court's decision-making was relatively straightforward. The verdict? Three years of jail time. Crucially, however, the possibility of this verdict depended on the anticipatory work done by various actors prior to this hearing. Working out the operational details and navigating uncertainties in the lab, as well as sensitising the court to a different 'way of seeing', anticipatory work here included operational and epistemic dimensions.

Conclusion

In the preceding pages, we have demonstrated how the 'adoption' of MPS in legal practice was predicated on a host of anticipatory practices involving the making of promises, the production of a shared epistemology within which MPS could be understood and, importantly, the tinkering and testing of the technology's operational aspects. In this sense, the Brettenpad case was a testing case not only for MPS but also for the analytic emphasis placed on epistemological and operational dimensions of the production of 'adoption spaces' as outlined by Wienroth (2018). Wienroth's analysis is helpful, indeed, as it has oriented our attention to moments when epistemological claims about its uses – it allows one to 'see more' or 'see things differently' – become crucial rhetorical devices in generating interest in this novel technology. At the same time, and bearing in mind Fortun's notes on the excessive character of the futures that are folded into technologies, we have aimed to highlight in particular the tinkering and even playful character of anticipatory work. In other words, we emphasise that such lab work is precisely the space in which possible futures and uses come to be carefully and playfully speculated on. Crucial here are also the role of the court and specific professional conceptions of judicial duties and tasks. While it asserted that it is not the actor to formally request MPS, it did go so far as to signal to the public prosecution that it would be receptive to an appeal in which MPS could then be used.

The extent to which MPS, following this rather eventful verdict, will become part of a broader forensic infrastructure remains, of course, to be seen. In some ways, MPS may always remain an anticipatory technology, not only because its application relies on the promise of clarification in cases involving complex DNA mixtures but also because, as a technology, it relies on an assessment of probabilities, and operationally, it requires invisible work on behalf of a host of actors. However, as a powerful 'promise machine', it has the capacity to significantly shape the technolegal scene. Promising more precision in notoriously difficult mixed profile cases, more frequent use in rape cases in particular is to be anticipated, as these tend to include mixed profiles with two or more donors. Such uses are likely not limited to the Dutch legal system. Here, the concept of the technolegal world is especially apt, as it combines within it an understanding of the 'worlding' that performatively takes place through specific technologies, as well as the novel possibilities for technolegal networks that come into being as a result. With MPS, criminal justice actors may come to learn to 'see differently', to distinguish more carefully between different donors and match profiles with more accuracy. But the adoption of MPS in this specific case has also created a precedent, not just nationally but globally, enabling forensic geneticists and lawyers across different jurisdictions not only recourse to accredited laboratories[15] but also access to the crucial legal technology of precedent. Indeed, while technologies may be methods for generating possibilities and futures, legal worlds tend to be focused on precedent and past legal decisions. In this sense, this specific case also offers an encounter with the multiple temporalities – past as well as future oriented – at play in technolegal worlds. The precise ways such temporalities intersect and inform the adoption of specific technologies is a valuable site of further study (but see Beynon-Jones and Grabham, 2019; Grabham et al., 2018; van Oorschot, 2014, 2020). Technolegally indeed, developments continue apace, with the recent approval of Cellmark's Abingdon laboratory as an accredited MPS lab in the UK.[16] The recent adoption of MPS, then, represents not only the translation of a promissory technology into an applied one but also the opening up of novel possibilities and futures for legal actors.

Acknowledgements

We are grateful to Victor Toom and Matthias Wienroth for their careful reading and helpful comments on draft versions of this chapter. We would also like to thank the technicians and researchers at the forensic laboratory for helping us understand MPS and the complexities of this case. Finally, we thank the ERC for supporting our research through an ERC-Consolidator Grant (fp7–617451-RaceFaceID-Race Matter: On the Absent Presence of Race in Forensic Identification).

Notes

1 Also referred to as next generation sequencing (NGS).
2 Verdict published by the Amsterdam court 13 April 2017 (emphasis added).
3 Verdict published by the Amsterdam court 28 February 2019 (emphasis added).
4 See www.wbur.org/news/2017/04/18/twin-dna-test-ruling (accessed 6 December 2020).
5 But also in cases involving identical twins. See, for example, www.bbc.com/news/magazine-25371014 (accessed 8 December 2020).
6 To be sure, his guilt is not merely based on DNA.
7 See also Cynthia Selin's work (2011) on establishing 'plausible futures'.
8 In the Dutch system, establishing a 'match' means that at least seven markers between two profiles correspond. The amount of loci on which this particular match was based remained unspecified in the judicial report and was also unknown to expert B.
9 Of course, the case file is not the sole means by which information about the case reaches judges; recent analyses have also pointed to the ways in which popular media became instrumental in shaping and co-creating cases; see, e.g., Jong and M'charek (2018). At the same time, the case file represents the procedurally bound and guarded collection of evidence to be judged and, consequently, the formal delineation of the case (e.g., van Oorschot and Schinkel, 2015).

10 The number of markers that are included in these analyses depends on the kit that is used.
11 Verdict published by the Amsterdam court 13 April 2017.
12 A sequencer is a scientific instrument that is used to record the order (sequence) of the four DNA bases in a given DNA sample.
13 The analyses of traces discussed here were not of the Brettenpad samples.
14 Polymorphisms of a single base occurring on the Y chromosome.
15 At the present moment, only a small number of forensic laboratories worldwide have introduced MPS technology as its implementation requires a significant turnover of laboratory infrastructure. This has discouraged labs from introducing it.
16 See https://verogen.com/cellmark-and-verogen-bring-next-gen-forensic-dna-to-uk/ (accessed (8 March 2020).

References

Beynon-Jones, S. and Grabham, E. (2019) *Law and Time*. London: Routledge.

Budowle, B., Onorato, A.J., Callaghan, T.F., Manna, A.D., Gross, A.M., Guerrieri, R.A., . . . McClure, D.L. (2009) Mixture Interpretation: Defining the Relevant Features for Guidelines for the Assessment of Mixed DNA Profiles in Forensic Casework. *Journal of Forensic Sciences* 54(4): 810–821.

Cole, S.A. and Lynch, M. (2006) The Social and Legal Construction of Suspects. *Annual Review of Law and Social Science* 2: 39–60.

De Knijff, P. (2019) From Next Generation Sequencing to Now Generation Sequencing in Forensics. *Forensic Science International: Genetics* (38): 175–180.

Fortun, M. (2005) For an Ethics of Promising, or: A Few Kind Words about James Watson. *New Genetics and Society* 24(2): 157–173.

Grabham, E., Cunliffe, E., Douglas, S., Keenan, S., Mawani, R. and M'charek, A. (2018) Exploring Relationships between Time, Law and Social Ordering: A Curated Conversation. *Feminist@Law*. https://doi.org/10.22024/UniKent/03/fal.652

Hopman, R. and M'charek, A. (2020) Facing the Unknown Suspect: Forensic DNA Phenotyping and the Oscillation between the Individual and the Collective. *BioSocieties* 15: 438–462.

Jeffreys, A.J., Wilson, V. and Thein, S.L. (1985) Hypervariable 'Minisatellite' Regions in Human DNA. *Nature* 314: 67–73.

Jobling, M.A. and Gill, P. (2004) Encoded Evidence: DNA in Forensic Analysis. *Nature Reviews Genetics* (5): 739–751.

Jong, L. and M'charek, A. (2018) The High-Profile Case as 'Fire Object': Following the Marianne Vaatstra Murder Case Through the Media. *Crime Media Culture* 14(3): 347–363.

Kruse, C. (2016) *The Social Life of Forensic Evidence*. Oakland: University of California Press.

Lander, E.S. (1989) DNA Fingerprinting on Trial. *Nature* 339(6225): 501–505.

Lynch, M., Cole, S.A., McNally, R. and Jordan, K. (2008) *Truth Machine: The Contentious History of DNA Fingerprinting*. Chicago: University of Chicago Press.

M'charek, A. (2000) Technologies of Population: Forensic DNA Testing Practices and the Making of Differences and Similarities. *Configurations* 8(1): 121–159.

M'charek, A. (2008) Silent Witness, Articulate Collective: DNA Evidence and the Inference of Visible Traits. *Bioethics* 22(9): 519–528.

M'charek, A. (2020) Tentacular Faces: Race and the Return of the Phenotype in Forensic Identification. *American Anthropologist* 122(2): 369–380.

M'charek, A., Hagendijk, R. and De Vries, W. (2013) Equal Before the Law: On the Machinery of Sameness in Forensic DNA Practice. *Science, Technology, & Human Values* 38(4): 542–565.

Mol, A., Moser, I. and Pols, J. (Eds.) (2010) Care: Putting Practice into Theory. 7–27. In *Care in Practice: On Tinkering in Clinics, Homes and Farms*. Bielefeld: Transcript Verlag.

Ossorio, P. (2006) About Face: Forensic Genetic Testing for Race and Visible Traits. *The Journal of Law, Medicine & Ethics* 34(2): 277–292.

Rheinberger, H-J. (1997) *Toward a History of Epistemic Things: Synthesizing Proteins in the Test Tube*. Stanford, CA: Stanford University Press.

Roewer, L. (2013) DNA Fingerprinting in Forensics: Past, Present, Future. *Investigative Genetics* 4(22).

Selin, C. (2011) Negotiating Plausibility: Intervening in the Future of Nanotechnology. *Science and Engineering Ethics* 17: 723–737.

Skinner, D. (2020) Forensic Genetics and the Prediction of Race: What is the Problem? *BioSocieties* 15: 329–349.

Star, S.L. and Strauss, A. (1999) Layers of Silence, Arenas of Voice: The Ecology of Visible and Invisible Work. *Computer Supported Cooperative Work (CSCW)* 8: 9–30.

Thompson, W.C. (1993) Evaluating the Admissibility of New Genetic Identification Tests: Lessons from the "DNA War". *The Journal of Criminal Law and Criminology* 84(1): 22–104.

Toom, V. (2011) *Dragers van Waarheid*. Amsterdam: Wolter Kluwers.

Tucker, T., Marra, M. and Friedman, J.M. (2009) Massively Parallel Sequencing: The Next Big Thing in Genetic Medicine. *The American Journal of Human Genetics* 85(2): 142–154.

Ulucanlar, S., Faulkner, A., Peirce, S. and Elwyn, G. (2013) Technology Identity: The Role of Sociotechnical Representations in the Adoption of Medical Devices. *Social Science & Medicine* 98: 95–105.

Van Oorschot, I. (2014) Het Dossier-in-Actie: Vouw-en Ontvouwpraktijken in Juridische Waarheidsvinding. *Sociologie* 10(3): 301–318.

Van Oorschot, I. (2018) *Ways of Case-Making*. Dissertation, Erasmus University Rotterdam.

Van Oorschot, I. (2020) Culture, Milieu, Phenotype: Articulating Race in Judicial Sense-Making Practices. *Social and Legal Studies* 29(6): 790–811.

Van Oorschot, I. (2021) *The Law Multiple: Judgment and Knowledge in Practice*. Cambridge: Cambridge University Press.

Van Oorschot, I. and Schinkel, W. (2015) The Legal Case File as Border Object: On Self-Reference and Other-Reference in Criminal Law. *Journal of Law and Society*, 42(4): 499–527.

Wienroth, M. (2018) Governing Anticipatory Technology Practices. Forensic DNA Phenotyping and the Forensic Genetics Community in Europe. *New Genetics and Society* 37(2): 137–152S.

Wienroth, M. (2020) Socio-Technical Disagreements as Ethical Fora: Parabon NanoLab's Forensic DNA Snapshot™ Service at the Intersection of Discourses Around Robust Science, Technology Validation, and Commerce. *BioSocieties* 15: 28–45.

Wienroth, M., Morling, N. and Williams, R. (2014) Technological Innovations in Forensic Genetics: Social, Legal and Ethical Aspects. *Recent Advances in DNA & Gene Sequences* (Formerly Recent Patents on DNA & Gene Sequences) 8(2): 98–103.

7

DELIBERATING FORENSIC GENETICS INNOVATIONS

The case of rapid DNA technologies in England and Wales

Dana Wilson-Kovacs

Introduction

Research on the value and utility of technology in policing in England and Wales, especially in relation to the 'practical application of different forms of scientific knowledge' (Innes et al. 2005: 39) is sparse (Williams and Weetman 2013). New forensic DNA technologies are a prime example: seldom examined in situ, their use has received scant sociological attention (Wilson-Kovacs 2014). This chapter aims to fill this gap through an analysis of the envisaged introduction of rapid DNA solutions in crime examination. In a changing landscape of forensic service delivery, marked by the dissolution of the Forensic Science Services (FSS) in 2012 and ongoing budget cuts to police forces, the promise of rapid DNA solutions brought together the Home Office, police personnel and commercial providers in an attempt to provide more efficient and cost-effective ways to process DNA trace. These new technologies were envisioned as a key driver in the implementation of scientific innovation in policing. They sought to produce DNA profiles from suspects and crime scene samples outside the laboratory environment within an hour and to speed the identification and elimination of suspects by submitting such profiles to the National DNA Database (NDNAD) for searches against individuals and/or crime scenes.

The chapter explores the deliberations surrounding the adoption of rapid DNA solutions and examines the ways in which their technological need has been articulated, justified and resisted. Together with the anticipated benefits surrounding the use of rapid DNA, the analysis explores the suitability, reliability, acceptability (McCartney et al. 2011) and potential impact of these technologies for investigative and criminal justice processes as understood by different key stakeholders, including those working for law enforcement agencies and the Home Office. The analysis draws on data collected between 2011 and 2018 through fifteen semi-structured

DOI: 10.4324/9780429322358-10

interviews with members of the National Policing Improvement Agency (NPIA), the Forensic Science Regulator Office and non-departmental advisory groups to the Home Office, as well as crime scene examiners and managers, senior investigating officers and other scientific support personnel affiliated with two police forces. Additionally, document analysis of the NPIA website material, the open minutes of the NDNAD Strategy Board and Ethics Group[1] meetings and the DNA Analysis Specialist Group was undertaken to provide supplementary information and corroborate findings from the interview data. Aiming to illuminate the different and changing logics, rationales and practicalities related to its adoption, the argument considers the utility of rapid DNA at the crime scene, in the custody suite and as evidence in court and considers its implications for faster investigations and courtroom evidential status.

The analysis builds on the need to go beyond the most visible aspects of a technology and pay attention to the normativities presented in discourses and materialities surrounding their use (Jasanoff 2003) and to document different understandings of the role of forensic DNA technologies (Prainsack and Toom 2013). It focuses on how arguments for the adoption of rapid DNA solutions capitalise on the tropes of efficiency, value and effectiveness. These tropes have characterised the UK's approach to forensic service provision since the launch of the DNA expansion programme in 2000 and the subsequent consolidation of the NDNAD (Skinner and Wienroth 2019). The argument illustrates how evolving critiques of the technology mobilise similar repertoires to scrutinise its need, utility and robustness. It considers how the interplay between the organisational drive for cost, speed and effectiveness on one hand and the production of accurate, credible and reliable evidence on the other render this attempt at technological adoption a conflicting undertaking. The contribution the chapter aims to make is twofold. First, to provide a timely update on how the collection and processing of DNA are devised as part of a discursive regime of efficiency, value for money, accountability and expertise. Secondly, the argument seeks to demonstrate the importance of a ground-up approach that draws on the experiences of the users of forensic technologies to explore how standardised forensic accomplishments occur (Williams and Weetman 2013).

The discussion proceeds as follows: the second section introduces rapid DNA technologies and the underlying rationale and framework for their intended adoption in England and Wales and discusses rapid DNA as an example of the police-led integration of forensic science into criminal investigation (Lawless 2011; Williams and Johnson 2008). The third section considers the suitability of rapid DNA solutions and outlines the discursive strategies and coexistent tensions as they emerge in the justification and contestation of new forensic DNA technologies. This part also reflects on the impact of their technological adoption on investigative practice, the criminal justice system and the forensic market. The conclusion calls for a situated and multi-perspectival understanding of the contexts in which the introduction of rapid DNA solutions has been envisaged. It highlights the need to consider how these technologies fit into existing organisational arrangements, where claims of their universality can be tested against locally curated technolegal ensembles as well as operational arrangements, policing routines and courtroom outcomes.

Setting the scene: Rapid DNA technologies

Rapid DNA technologies are commercially provided tools that can produce a DNA profile faster than can be done using conventional techniques. They involve the fully automated separation and detection of DNA material, its extraction and amplification from swabs taken at crime scenes or in custody and provide test results in under two hours. Rapid DNA solutions can be used for screening DNA traces at crime scenes, profiling DNA samples in custody suites, paternity testing and identification at mass fatalities sites (Geldenhuys 2018; Grover et al. 2017; Murphy 2018; Turingan et al. 2016). The United States established a rapid DNA programme in 2010 to explore the opportunities presented by the potential of this technology for DNA laboratory testing backlogs and law enforcement agencies, especially border control. Today, rapid DNA technologies have been adapted worldwide for the examination of major crime.[2] Two main types of technologies are used: one for profiling traces of DNA and connecting those with existing database profiles of suspects and the second for screening DNA traces to determine their provenience and whether a full profile can be obtained by submitting the sample for laboratory analysis. Rapid DNA technological solutions integrate DNA extraction and PCR amplification of 15 or more STR loci. Because they typically use different profiling chemistries and a smaller number of STR loci than profiles currently required for the NDNAD, the retention of profiles obtained through rapid DNA solutions on the national database has been problematic. Consequently, portability, speed of identification and ease of use have been emphasised as key features in the marketing of rapid DNA solutions in a linear discourse that builds on stories of forensic DNA profiling as the most successful technologies of solving crime. As a result, it echoes similar narratives that highlight 'DNA as the triumphant hero' (Aronson 2007; Lynch et al. 2008; M'charek 2008). The envisaged uptake of rapid DNA technologies reflects the anticipatory practices related to other forensic innovations, such as DNA phenotyping (Wienroth 2018; M'charek et al. 2020).

In the UK, the rationale accompanying the rapid DNA adoption aligned to the commitment to technological innovation expressed in the Science and Technology Strategy 2009–2012 (Home Office 2009). Key agencies such as NPIA (2007–2012) and the Home Office Centre for Applied Science and Technology (2013–2018) provided the organisational infrastructure to support the development of rapid DNA technologies, initially through the Forensics21 programme. Launched in 2007, Forensics21 sought to identify 'how forensic science may be used to best effect' to provide direct benefits to policing, such as 'reduced investigation time, increased detections, increased "cold hits"; shorter time to arrest; improved public confidence; new links to outstanding crime; [and] early intervention' (Bramble 2009: 4). Bringing together commercial providers and representatives of the FSS and the Home Office, Forensics21 focused on increasing accountability and improving quality and cost benefits. It linked public sector priorities to technological developments to support the standardisation, optimisation and monitoring of forensic support delivery across police forces.

In the case of rapid DNA technologies, the programme monitored their development and adjustment to policing needs through the Accelerated DNA Profiling Technologies (ADAPT) initiative. Overseen by members of the DNA Analysis Specialist Group and NDNAD Ethics Group, ADAPT was designed to rationalise and improve forensic delivery through the introduction of portable tools that can produce full and accurate DNA analyses of crime traces outside a laboratory environment within an hour. The initial call for rapid DNA solutions launched in October 2009. It requested technologies that 1) reduce the time taken between the retrieval of DNA material, 2) obtain a match with individual profiles held on the NDNAD and 3) maximise the ability to search a detainee's DNA profile against unsolved crime stains. However, following collaborative 'proof of concept' pilots between commercial developers and police forces to adapt the technology to policing needs in 2010, expectations were adjusted. Consequently, screening solutions (i.e., checking at the crime scene whether a trace contains enough human DNA for laboratory analysis) rather than profiling ones (i.e., matching crime scenes profiles with those on the NDNAD) were identified as suitable.

In 2010, a three-phase approach was crystallised, with economies calculated at each step. This aligned prototypes with the suppliers' development timescales and presented more clearly defined cost benefits. First-phase technologies sought the elimination of samples containing no detectable DNA before submission to a forensic science provider. This step was estimated to save forces around £3 million per annum (ADAPT Supplier Meeting notes, December 2011). Second-phase solutions would employ more advanced techniques to indicate whether a sample is likely to provide a full DNA profile and would be suitable to load onto the NDNAD. Additionally, the use of partial profiles from viable samples for initial intelligence on detainees and suspects was also identified as a possible benefit. Altogether, the projected savings to forces were £8 million per annum (idem). While third-phase solutions were closest to the profiling technology originally envisaged to give DNA profiles fully compatible with the NDNAD within an hour, it was now recognised that they would not be available in the short to medium term. Adjustments in the scale and scope of the initially envisaged technologies were also evident in the reduction of the types of environments where third-phase solutions could be used to only one – i.e., the custody suite – to allow for the control of variables and operator errors.

After the dissolution of NPIA in 2012, ADAPT continued as part of the Home Office Centre for Applied Science and Technology. Shifting from claims for an imminent change in operational policing and the supply of information to investigating officers, its aim became to identify possible applications and the infrastructure needed for rapid DNA solutions, including satisfying the courts of its validity and assessing its potential impact on the forensics marketplace. Subsequently, rapid DNA pilots have been undertaken by a few forces to test the suitability of prototypes for crime scenes and help providers narrow the gap between expectations and delivery. However, progress has been slow. In its 2016 Forensic Science Strategy, the Home Office highlight the need to ensure 'that lessons have been learned from field trials of technology like "Rapid DNA"' (Home Office 2016: 26). Notes of the March 2018 meeting of the Biometrics and Forensics Ethics Group mention the group's input to the ongoing piloting of rapid DNA technologies, and minutes of the November 2019 meeting of the DNA Analysis Special Group (DNASG) Strategy Board highlight discussions about the suitability of various rapid DNA capabilities.

In a socio-political landscape dominated by fixed organisational priorities and neo-liberal market discourses (Lawless and Williams 2010), the introduction of rapid DNA technologies was envisaged to bypass laboratory demands and waiting times by providing an out-of-the-box solution that required little or no forensic expertise. It unfolded as part of a drive for economy, efficiency and accountability, in which cost savings, the speed of identification and the accuracy of hits addressed shortcomings in the streamlining and processing of DNA trace. Elsewhere, commentators noted how the introduction of rapid DNA solutions might result in the reduction of the number of items requiring analysis and offer a 'Model "T" Ford production line approach on the number of items and cases that yield robust evidence' (Raymond 2015: 368).

In the UK, the initiative further instanced the reproduction of a 'relatively unproblematic' rationalisation and police-led view of forensic provision (Lawless 2011: 675). Alongside the government-led DNA Expansion Programme (1999–2004) and the development of the NDNAD, it provides another example of a rationalisation that tailors the application of forensic science to police objectives and needs (Lawless 2011). While for major and serious crime, this implies the alignment of appropriate expertise with the specifics of particular cases, for volume crime, 'rationalisation relates to the development of robust systems, rules, habits and standards that can be rolled out and routinely applied across many cases maximising the net benefit at minimum costs' (Tilley and Townsley 2009: 360). Yet the rationalisation of forensic provision in volume crime has proven hard to achieve, with government reviews outlining persistent issues in the management of forensic science resources (Lawless 2016). The introduction of rapid DNA solutions has aimed to address some of these issues through economic measures, such as delimiting the number of items needing external laboratory analysis and reducing the cost of expertise required. In order to understand the ongoing process of the rationalisation of forensic provision, a consideration of its specific contexts of application and of economic, professional and regulatory parameters is key. Illustrating this, the following section explores the stakeholders' explanations of the place and role of rapid DNA solutions in police work in the UK. The analysis contrasts NPIA accounts that support an understanding of these developments as tools for the speedy identification of suspects and the triage of exhibits for DNA trace before submitting samples for laboratory analysis with users and regulators' perspectives that underline the wider challenges these technologies foretell. In so doing, it offers an empirical insight into the technolegal world of emerging rapid DNA technologies envisaged for everyday use in criminal investigations: i.e., the different discursive practices and values that legitimise their development.

Stakeholders' narratives

1. Legitimising rapid DNA solutions

The accounts of the NPIA representatives helped contextualise the introduction of rapid DNA solutions, situating it in a wider process of forensic innovation on the lines established by the FSS. These participants presented the broad political and institutional parameters for the legitimisation of rapid DNA solutions and located the impetus for these tools in the US, where backlogs of unprocessed evidence, border control problems and the readiness of commercial laboratories to address these issues configured the business requirements (Lovrich et al. 2004). In contrast, its introduction in the UK was discussed in a context of innovation and accessibility, rather than immediate utility, as one NPIA official explained:

> We're trying to speed up . . . applications in policing as [technology] gets to be more accessible in a box. . . . [W]e could wait until that was done and then, 'oh, look, we could use it in policing'. Well, no, we want, as it's moving in that direction, we're trying to intervene to get the technology into the police to help society at an earlier stage than it otherwise may have done.
>
> *(03, NPIA)*

Ongoing engagement with police users was seen to channel development through pilot trials, allow for early intervention and tailor products to force requirements. This process typically started with helping the users to 'understand better what the potential is':

> We're trying to bring the customer requirement much nearer the beginning of that innovation thinking rather than [having] a product ready built. . . . [It] enables the customer . . . the police, to understand better what the potential is. So . . . through the trials, the customer might understand more about how this could be beneficial.
>
> *(01, NPIA)*

Convincing users of such potential was framed unproblematically, and rapid DNA technology was seen as providing unprecedented speed of identification, improvement in detection and cost savings to forces. The economic imperative of the latter was described as the key driver for early adoption: '[W]e can save lots of money, millions of pounds a year by not sending samples that actually wouldn't give us a result' (01, NPIA).

> The whole police architecture is changing by financial constraints. . . . We may end up with DNA clearing houses, instead of each force having a centralised DNA process we may have just one for the whole region. There'll be a lot more emphasis on the business case around what it saves in money,

> rather than what it delivers in detection which previously it may have been more of a persuader.
>
> *(02, NPIA)*

Building on the tradition of FSS forensic innovation, a vocabulary of societal need, technological speed, reduced costs and procedural efficiencies is prominent in these accounts. When asked about the potential challenges in implementing the new technologies, NPIA stakeholders identified these as over-optimistic delivery timelines from commercial providers, less flexible infrastructural policing provision and police resistance to changing custody suite routines. Technical matters such as the size and complexity of some of the samples that would require specific laboratory analysis are further acknowledged as potential challenges. Overall, however, rapid DNA is presented as advantageous, fuelling the expectation of an automated, live system in which links between sample submissions and the NDNAD are (semi) instantaneous. Furthermore, the technology is described as beneficial for the original envisaged users – the scene of crime officers (SOCO) – for whom the change is presented as empowering, giving them a sense of 'owing the job' and enabling self-esteem. These views reinforce the promissory discourse of rapid DNA as a relatively autonomous external driver of organisational transformation (Orlikowski 2009), in which DNA trace submissions are streamlined to external forensic service providers, unsuitable samples are eliminated and the technology provides the step change to ensuring value for money and efficiency at force level (measured in faster identification times and improved hit rates).

2. Challenging the business case

While these NPIA accounts illuminate rhetorical strategies through which forensic DNA developments are made to count for effective policing, stakeholders involved in the governance of rapid DNA solutions focused on the impact of the technology in a criminal justice system context, bringing to fore unresolved issues of accountability, validation and reliability. Here, members of non-departmental bodies and advisory groups offering independent oversight expressed their doubts. Their concerns regarded the promise of the new technology in terms of the manufacturers' ability to provide the products,[3] the rush into development and testing and the robustness of the business case. These respondents compared the promised technological utility of rapid DNA solutions to other forms of crime control and reduction interventions that might yield better results less expensively:

> We simply don't know whether the investment in that sort of technology is going to achieve the results that we could achieve elsewhere by doing something different.
>
> *(04, DNA Database Ethics Group)*

Here, the predominant focus on value for money is no longer presented as deliverable through the technology, and costs are contrasted to other available crime control and surveillance technologies, such as CCTV. A reoccurring theme in these accounts is the lack of systematic evidence linking incurred expenditure to the effectiveness of forensic DNA technologies, which reflects well-documented related concerns (e.g., Ludwig 2016; Webb et al. 2005; Wilson et al. 2011). These accounts voice a more cautious approach to the introduction of rapid DNA solutions, with commercial assurances and potential savings to forces balanced against unrepairable damage to the credibility of judicial outcomes.

> In essence you're losing even more control of your DNA processing, that's going to third parties and you've got to be really robust on your process changes.
>
> *(07, NDNAD custodian)*

These stakeholders emphasise the need for a risk management strategy and impact assessment to address validation, accountability, the reputational costs for the criminal justice system and the admissibility of evidence obtained through the new technology in bringing cases to court[4]:

> [There's] a lot of testing and validation to be done yet. . . . We have a bespoke service running here with very demanding quality standards which include proficiency testing . . . accreditations . . . audit checks, quality checks . . . a whole range of things going on. If we move to this new . . . process, I still don't know who's going to be accountable for quality and I am far from convinced, and I'm waiting to be convinced, that we can maintain that gold standard, so there's a big question mark about the reliability and integrity of the product we're going to get through the speeding-up process.
>
> *(06, Forensic Science Regulation Unit)*

As forces become more strategic in the selection of submissions sent to external forensic laboratories and their on-site facilities undergo accreditation to ISO17025 standards to process DNA trace, the adoption of rapid DNA solutions is seen as potentially destabilising both the hard-earned reputation of the 'bespoke service' and the health of independent forensic science provision laboratories. While rapid DNA may help reduce the costs of sending exhibits to these laboratories, the solutions are also instrumental to the shaping of forensic demand. Here, participants feared that the decrease in the number of samples submitted to external providers might impact negatively on the size of the forensic market:

> If you go for a [rapid DNA] type solution, you, in effect, close a lot of this [market] down. You still need to keep some laboratory functions there for confirmatory work, but you're taking probably, at a very rough guess, £40 million out of that market. What does that do for the stability of a forensic science provider market in the future?
>
> *(05, Forensic Science Regulation Unit)*

This question captures the uncertainty caused by the closure of the FSS and the potential further impact the introduction of rapid DNA technologies may have on a diminishing number of independent forensic service providers. This problem is compounded by the difficulties in assessing forensic expenditure nationally (Ludwig 2016). Given the lack of standardised financial systems across police forces and agreement over what constitutes forensic spending, local costs can be interpreted and recorded differently, which hinders both comparisons across forces[5] and assessing the size of the UK forensic services market. Relatedly, an evaluation of the forensic market, using publicly available data, identified the overall forensic expenditure through the National Forensic Framework, which assists police forces in purchasing services from private forensic service providers at a significant decline in comparison with previous years (House of Commons Science and Technology Committee Briefing, December 2014).

3. *Using rapid DNA technologies*

Embedded in a day-to-day understanding of the collection and use of DNA trace in relation to other forensic techniques, local infrastructures and organisational routines, potential users interrogated the feasibility of these technologies for both volume crime investigation and custody-based settings. Typically questioned was the need for faster suspect identification once in custody. Considering that taking fingerprints and DNA from suspects when first detained is mandatory, the added value of rapid DNA to extant procedures was difficult to gauge:

> If we identify that you're not on the database but we think you've been committing a number of rapes around here or burglaries, we want to know your DNA profile before we release you from custody. That's when you want your Rapid DNA. But those are quite rare occasions, because the majority of our criminal population . . . are already on the DNA database.
>
> *(11, Forensic Science Manager)*

As DNA hits matter in terms of the economic and investigative value of the submissions and the performance of forces, making careful decisions regarding the selection of potential samples is key:

> We've got a converter team who deal with our DNA and fingerprint hits when they go out and investigate. But we're now talking about doing that in rapid time, 24-hour round the clock service. Forces are cutting back, we're losing . . . police officers, civilian investigators, posts are going across forces. . . . How can we have the manpower to go out and invest all the time and money in a DNA hit from what? Theft from a motor vehicle with a bit of blood on. Are you going to go out and nick that offender straight away? What resources are you going to chuck at it?
>
> *(09, Crime Scene Manager)*

Reservations regarding new forensic DNA technologies were also expressed in relation to the increase in immediate costs to forces. A discursive frame of cost effectiveness was employed in these accounts to justify current arrangements, despite the promise of long-term savings. Although the use of technology in certain scenarios is acknowledged, additional concerns linked to competency and the credibility of expertise within the criminal justice system procedures echo those posed by stakeholders involved in the governance of forensic technologies:

> Perhaps there is some benefit there, that you could actually have a bloodstain, or a semen stain . . . and you could bring a piece of kit into that scene. . . . I question the logistics of the whole thing, how it's going to work and to have staff proficient and able and happy to be the person who produces a DNA profile on a bit of kit that they do not understand and couldn't (explain). . . standing in the witness box, 'how does this equipment work?' 'I don't know, I just push the button and it gives me the profile'.
>
> *(14, Scene of Crime Officer)*

Here, the introduction of rapid DNA technologies is interrogated using repertoires of efficiency and accountability mobilised to reinforce the value of extant practices, administrative processes, submission times and workloads. Moreover, echoing the doubt of rank-and-file police officers towards new technologies (Chan 2001; Manning 2008), users' perspectives illuminate the professional dimensions accompanying the introduction of rapid DNA solutions. In contrast with NPIA views on rapid DNA as potentially empowering forensic support personnel, users regarded rapid DNA solutions with scepticism. Similar to the collection of trace for DNA profiling, they saw them as among the more menial tasks perceived to take away from rather than add to their professional status and expertise (Wilson-Kovacs 2014). Views of encroaching technology of questionable practical value, described as 'everything that your Robocop SOCO[6] would love to have in his tool bag' (13, Technician, Scientific Support Unit) accompanied these reservations:

> We're being driven by technology and by the demand for a speedier, faster service in everything we do. . . . The FSS went down that line with lab-in-a-van, because they thought that it'd be great to develop a profile from a van parked outside a major crime scene. There and then, beam it off on a satellite to the database, search the database and get a result back. . . . Your offender is 'blah, blah, blah' . . . but. 'well, hang on, by the time it takes a laboratory to get out with the van, all the staff on board, all cleaned up, all sterilised, the van cleaned out from the last time, takes five hours to drive down here in the middle of the night, you have to get people out of bed, we could have traffic motorcyclists take that sample that we've found at the crime scene to the lab ready.
>
> *(12, Scene of Crime Officer)*

These remarks reinforce the contradictory picture offered by analyses of the impact of technology on daily policing tasks (Kobus et al. 2011) and add to extant literature on the perspectives of scientific support and crime scene personnel. They illustrate the problematic nature of the ways in which top-down initiatives have been envisaged to streamline the cost of submissions and boost efficiency and effectiveness. These situated perspectives outline the gap between the imposed implementation of forensic technologies and their perceived utility at the local level. They highlight the potential disruption to organisational routines and challenges to professional expertise that rapid DNA solutions are anticipated to bring.

More generally, not only does the introduction of rapid DNA reflect the uncertainties of a forensic landscape in flux and the changing dimensions of forensic provision but its trajectory also captures the tensions between enduring visions of forensic promise and their practical implementation. In their legitimisation of activities that aim to reduce uncertainty about the future, these visions resemble accounts of innovation in regenerative medicine (e.g., Brown et al. 2000; Borup et al. 2006; Erikainen and Chan 2019). The promissory expectations of rapid DNA also have a performative character: as reviewed specifications fine-tune envisaged efficiencies in forensic delivery, economic costs and public accountability, the crystallisation of rapid DNA solutions comes 'to occupy the full space of possibility' and enables 'rhetorical closure' (Garb 2005; Bloor et al. 2014: 241). The perspectives presented here illustrate distinct vocabularies of motives and specific group interests. They illuminate the contexts and negotiations surrounding the development and introduction of these technologies in policing, the benefits and risks they carry and their place in existing routines. While participants acknowledged rapid DNA solutions as aids in crime detection, the challenges posed by their envisaged implementation call for a careful consideration of the issues raised by the development and adoption of these technologies before such rhetorical closure is achieved.

The stakeholders' perspectives overlap in their cautionary approach towards manufacturers' claims and ability to deliver to the specifications required by the Home Office call for technological solutions.[7] They also highlight the lack of financial incentives for forces and the capital investment needed to adopt rapid DNA solutions. Unlike the centralised, supported investment in the NDNAD and the DNA Expansion Programme, the adoption of rapid DNA is optional to forces and subject to local budget demands. The views presented diverge in the emphasis given to particular aspects in the process of development and adoption. Both users and regulators question the acceptability of rapid DNA evidence in court, with the latter challenging the business case, emphasising the potential for miscarriages of justice and reputational loss and interrogating the impact of rapid DNA on a diminishing forensic market. Views are divided between stakeholders embracing the change and the opportunities offered through various commercial providers and those cautioning of limitations.

This ambivalence is captured in the minutes of the NDNAD Ethics Group and those of the National DNA Strategy Board. As the rapid DNA Technology Project continued, so did the scrutiny of rapid DNA solutions and the proposed pilots to test them at crime scenes and in custody suites. While one force purchased one of the screening solutions to help them decide whether a sample should be submitted to a forensic provider for full analysis, the Metropolitan Police reported discontinuing the same solution for 'valid reasons'. The open minutes from September 2014 (Minutes of the National DNA Strategy Board Meeting, 2014), for instance, note continuing concerns with the business benefit and model driving the development of the technology, with members stating that manufacturers have still to demonstrate that 'the science was good enough'. Critically, also noted is the ongoing need for the Home Office input into the scientific decisions surrounding the suitability of the proposed solutions. Rather than taking manufacturers' claim 'at face value', it is noted that they need to be questioned 'in view of evidence of the reliability of solutions offered by the Forensic Science Regulator and the Metropolitan Police' (8.10, p. 10).

The social, reputational and operational issues linked to the development of rapid DNA technologies reflect distinct views that are collectively shared and strategically mobilised in the debate as the respondents' accounts presented here and the cited minutes of the National DNA Database Strategy Board and the NDNAD Ethics Groups attest. Notably, the minutes capture how these vocabularies are employed to prolong debate; question the applicability, efficiency and cost effectiveness of rapid DNA solutions; and tease out the longer-term, less acknowledged risks and implications accompanying the promissory claims regarding the speed of identification and savings to forces. Similar to discussions of the social, legal and ethical aspects of new forensic technologies (Wienroth et al. 2014; Williams and Wienroth 2017), they reinforce stakeholders' considerations on the testing, validation and governance of new forensic DNA technologies, the contexts of their application and their use in the criminal justice system. Deliberations surrounding the implementation of rapid DNA technologies in England and Wales continue. A recently established rapid DNA working group[8] and a guidance document on methods for rapid DNA devices, published in April 2021 by the Forensic Science Regulator, demonstrate the continuing interest in these technologies (Forensic Science Regulator Guidance, 2021).

Concluding remarks

Rapid DNA has been presented as an important technological development in forensic science, with potential major consequences for criminal justice processes, the law's ability to dissect and question the credibility and epistemic authority of evidence and the public understanding of and engagement with forensic science. This chapter has considered deliberations surrounding the introduction of new forensic DNA technologies in policing in England and Wales with the view to optimising the collection and processing of exhibits from the perspective of multiple stakeholders. Challenging the linear vision of rapid DNA technologies offered by developers and government stakeholders, it provides an insight into how technological innovation is appropriated and re-imagined in different operational contexts. Whereas the initial scope of rapid DNA technologies as developed in the USA was to monitor border control and help with victim identification and laboratory backlogs, their adoption in the UK was justified through their benefits to solving volume crime. However, this justification has been doubted by some police users, who largely questioned the technology's potential for a new way of carrying out investigations. Moreover, the suitability of the technology for the courtroom and its implications for the forensic service marketplace have raised further questions on the quest to adopt rapid DNA solutions. Furthermore, the speed of miniaturisation chemistries, the complexity of technical and quality standards required and the operational arrangements and business changes envisaged have also been instrumental in adjusting the expectations surrounding this adoption.

The analysis introduced the operational framework and infrastructural provision envisaged to support this process and presented the overarching rationale, discursive strategies and coexistent tensions in the justification and critique of these technologies, as raised by those interviewed. It highlighted the uncertainty surrounding the police introduction of rapid DNA solutions and illustrated how this process is open to multiple understandings regarding the utility and reliability of these technologies by their potential end users and by those involved in their future governance. Rendering new technolegal worlds, it showed how strategic political decisions governed by efficiency and cost-saving rationales continue to configure the use of forensic technologies in policing and enable the commercialisation of forensic provision. Prima facie, the viewpoints presented here align with other value-for-money perspectives on forensic knowledge and the application of new forensic DNA technologies (Lawless and Williams 2010). They also illustrate how ambiguities around the investigative value and use of these technologies are expressed by various stakeholders. On one hand, repertoires of speed, efficiency and economy are used to frame technological introduction and to sustain a forensic imaginary (Williams and Weetman 2013) related to unprecedented rates of trace identification, long-term cost-savings to forces and maximisation of the NDNAD. On the other hand, the interplay between value, risk, uncertainty and reputational loss is used to question the need for rapid DNA technologies in volume crime, their scientific compatibility with existing systems of profile monitoring and their reliability in the courtroom.

The discussion illustrates how rapid DNA solutions consolidate the understanding of forensic science applications in policing in England and Wales as framed by cost savings and efficiency. This vision demonstrates a continuing, rationalised approach to the production of evidence within the strictures of new public management and neo-liberal discourses established with the DNA expansion programme and the development of the NDNAD (McCartney 2006). Reflecting such exigencies, the drive for rapid DNA solutions crystallises their role in the more accurate selection of exhibits for laboratory analysis (and therefore less wasteful) and the improvement in hit rates. Not only does the promotion of rapid DNA solutions encapsulate a view of forensic genetics innovation in policing as linearly developed in relation to the needs of its users and the pursuit of successful criminal justice outcomes, but this vision also consolidates the place of new DNA technologies as technical support subsumed to police priorities. Dominating debates on their operational utility, the promise surrounding the value and efficiency of these technologies counteracts concerns over their validity and reliability while bringing technological possibilities closer to fruition. It is hoped that the empirical contribution made in this chapter illuminates the complex relationship between forensic technologies and policing. As overarching political, regulatory and economic interests shape the scope and role of new forensic DNA technologies, the discussion presented sought to move the focus from one of technological hype and promise to one that places such developments in the context of their governance and use while also accounting for the inherent tensions in this process.

Notes

1 The National DNA Database Strategy Board provides governance and oversight over the operation of the National DNA Database (NDNAD) and the National Fingerprint Database. It includes representatives of the National Police Chiefs' Council (NPCC) and the Association of Police and Crime Commissioners (APCC), the Forensic Science Regulator, the Information Commissioner's Office, the Biometrics Commissioner, the Home Office and the DNA Ethics Group. The DNA Analysis Specialist Group and DNA Database Ethics Group provide scientific, technical, legal and ethical oversight and input into the NDNAD Strategy Board and are constituted by representatives of law enforcement agencies, the criminal justice system and academia.
2 For an understanding of the classification used, see https://www.met.police.uk/sd/stats-and-data/.
3 The tendency of suppliers to over-market and over-sell is often noted in official documents, such as in the Second Report of the House of Commons Science and Technology Committee (2013).
4 These concerns have continued to dominate rapid DNA pilots: the Minutes of the DNA Analysis Specialist Group meeting (December 2014) noted issues with proficiency, quality, and batch testing of kit from suppliers, alongside competency and instrument monitoring in the development of the technology.
5 Figures on forensic spending are incomplete, inconsistent and difficult to interpret: to provide a clearer picture of annual expenditure on forensics across all police forces in England and Wales, an estimated 372 documents would need to be examined (House of Commons Science and Technology Committee Briefing, December 2014).
6 Scene of crime officer (another title for crime scene investigators).
7 See for instance, the minutes of the National DNA Strategy Board, September 2014, section 8.
8 The minutes of the DNA Analysis Special Group (DNASG) Strategy Board Meeting, November 2019, section 5.

References

Aronson, J., 2007. *Genetic Witness: Science, Law, and Controversy in the Marketing of DNA Profiling*. New Brunswick, NJ: Rutgers University Press.

Bloor, M., Sampson, H., Baker, S. and Dahlgren, K., 2014. The instrumental use of technical doubts: Technological controversies, investment decisions and air pollution controls in the global shipping industry. *Science and Public Policy*, *41*(2), 234–244.

Borup, M., Brown, N., Konrad, K. and Van Lente, H., 2006. The sociology of expectations in science and technology. *Technology Analysis & Strategic Management*, *18*(3/4), 285–298.

Bramble, S., 2009. Forensics21: Challenging, enabling and improving forensic science. *Police Professional*, 8 January 2009. Accessed 07/06/2015 at www.policeprofessional.com/news.aspx?id=7907.

Brown, N., Rappert, B. and Webster, A., 2000. *Contested Futures: A Sociology of Prospective Techno-science*. Aldershot: Ashgate.

Chan, J.B., 2001. The technological game: How information technology is transforming police practice. *Criminal Justice, 1*(2), 139–159.

Erikainen, S. and Chan, S., 2019. Contested futures: Envisioning "personalized", "stratified," and "precision" medicine. *New Genetics and Society, 38*(3), 308–330.

Forensic Science Regulator Guidance, 2021, Methods employing rapid DNA devices. Issue 1. FSR-G-229: Crown Copyright.

Garb, Y., 2005. Constructing the trans-Israel highway's inevitability. *Israel Studies, 9*, 180–217.

Geldenhuys, K., 2018. The Kroonstad train tragedy-how Rapid DNA technology was used to bring closure to families. *Servamus Community-based Safety and Security Magazine, 111*(9), 32–35.

Grover, R., Jiang, H., Turingan, R.S., French, J.L., Tan, E. and Selden, R.F., 2017. FlexPlex27—highly multiplexed rapid DNA identification for law enforcement, kinship, and military applications. *International Journal of Legal Medicine, 131*(6), 1489–1501.

Home Office, 2009. *Science and Innovation Strategy 2009–2012*. London: HMSO.

Home Office, 2016. *Forensic Science Strategy*. London: HMSO.

House of Commons Science and Technology Committee, 2013 July. *Forensic Science. Second Report of Session 2013–14*. Volume I. HC610. London: The Stationery Office Limited.

House of Commons Science and Technology Committee Briefing, 2014 December. The home office's oversight of forensic services. Accessed 13/09/2016 at www.nao.org.uk/report/home-offices-oversight-forensic-services.

Innes, M., Fielding, N. and Cope, N., 2005. 'The Appliance of Science?' The theory and practice of crime intelligence analysis. *British Journal of Criminology, 45*(1), 39–57.

Jasanoff, 2003. Technologies of humility: Citizen participation in governing science. *Minerva, 41*, 223–244.

Kobus, H., Houck, M., Speaker, P., Riley, R. and Witt, T., 2011. Managing performance in the forensic sciences: Expectations in light of limited budgets. *Forensic Science Policy & Management: An International Journal, 2*(1), 36–43.

Lawless, C., 2011. Policing markets: The contested shaping of neo-liberal forensic science. *British Journal of Criminology, 51*(4), 671–689.

Lawless, C., 2016. *Forensic Science: A Sociological Introduction*. London: Routledge.

Lawless, C. and Williams, R., 2010. Helping with inquiries, or helping with profit? The trials and tribulations of a technology of forensic reasoning. *Social Studies of Science, 40*(5), 731–755.

Lovrich, N.P., Pratt, T.C., Gaffney, M.J., Johnson, C.L., Asplen, C.H. and Hurst, L.H., 2004. *National Forensic DNA Study Report*. Pullman: Washington State University. Accessed 08/09/2015 at www.ncjrs.gov/pdffiles1/nij/grants/203970.pdf.

Ludwig, A., 2016. E'value'ating forensic science. *Forensic Science Policy & Management: An International Journal*, 7(3/4), 66–77.

Lynch, M., Cole, S.A., McNally, R. and Jordan, K., 2008. *Truth Machine: The Contentious History of DNA Evidence*. Chicago: University of Chicago Press.

M'charek, A., 2008. Silent witness, articulate collective: DNA evidence and the inference of visible traits. *Bioethics, 22*(9), 519–528.

M'charek, A., Toom, V. and Jong, L., 2020. The trouble with race in forensic identification. *Science, Technology & Human Values*. https://doi.org/10.1177/0162243919899467

Manning, P., 2008. *Technology of Policing: Crime Mapping, Information Technology and the Rationality of Crime Control*. New York: New York University Press.

McCartney, C.I., 2006. The DNA expansion programme and criminal investigation. *British Journal of Criminology, 46*(2), 175–192.

McCartney, C.I., Wilson, T.J. and Williams, R., 2011. Transnational exchange of forensic DNA: Viability, legitimacy, and acceptability. *European Journal on Criminal Policy and Research*, *17*(4), 305–322.

Minutes of the DNA Analysis Specialist Group Meeting, 2014 December. Accessed 13/09/2016 at www.gov.uk/government/ . . . /DNASG_2014_12_11__minutes_vfinal.pdf.

Minutes of the DNA Analysis Special Group (DNASG) Strategy Board Meeting, 2019 November. Accessed 20/3/2021 at www.gov.uk/government/groups/national-dna-database-strategy-board.

Minutes of the National DNA Strategy Board Meeting, 2014 September. Accessed 15/09/2016 at www.gov.uk/government/groups/national-dna-database-strategy-board.

Murphy, E., 2018. Forensic DNA typing. *Annual Review of Criminology*, *1*, 497–515.

Orlikowski, W.J., 2009. The sociomateriality of organisational life: Considering technology in management research. *Cambridge Journal of Economics*, *34*(1), 125–141.

Prainsack, B. and Toom, V., 2013. Performing the Union: The Prüm decision and the European dream. *Studies in History and Philosophy of Science Part C: Studies in History and Philosophy of Biological and Biomedical Sciences*, *44*(1), 71–79.

Raymond, T., 2015. The musings of a forensic scientist retiree. *Australian Journal of Forensic Sciences*, *47*(4), 365–370.

Skinner, D. and Wienroth, M., 2019. Was this an ending? The destruction of samples and deletion of records from the UK police national DNA database. *BJHS Themes*, *4*, 99–121.

Tilley, N. and Townsley, M., 2009. Forensic science in UK policing: Strategies, tactics and effectiveness. In Fraser, J. and Williams, R. (eds.) *Handbook of Forensic Science*. London: Routledge. 359–379.

Turingan, R.S., Vasantgadkar, S., Palombo, L., Hogan, C., Jiang, H., Tan, E. and Selden, R.F., 2016. Rapid DNA analysis for automated processing and interpretation of low DNA content samples. *Investigative Genetics*, 7(1), 1–12.

Webb, B., Smith, C., Brock, A. and Townsley, M., 2005. DNA fast tracking. In M. Smith and Tilley, N. (eds) *Crime Science: New Approaches to Preventing and Detecting Crime*. London: Routledge. 167–186.

Wienroth, M., 2018. Governing anticipatory technology practices. Forensic DNA phenotyping and the forensic genetics community in Europe. *New Genetics and Society*, *37*(2), 137–152.

Wienroth, M., Morling, N. and Williams, R., 2014. Technological innovations in forensic genetics: Social, legal and ethical aspects. *Recent Advances in DNA & Gene Sequences*, *8*(2), 98–103.

Williams, R. and Johnson, P., 2008. *Genetic Policing: The Use of DNA in Criminal Investigations*. London: Willan Publishing.

Williams, R. and Weetman, J., 2013. Enacting forensics in homicide investigations. *Policing and Society*, *23*(3), 376–389.

Williams, R. and Wienroth, M., 2017. Social and ethical aspects of forensic genetics: A critical review. *Forensic Science Review*, *29*(2), 145–169.

Wilson, D., Weisburd, D. and McClure, D., 2011. Use of DNA testing in police investigative work for increasing offender identification, arrest, conviction, and case clearance: A systematic review. Accessed 01/11/2015 at www.campbellcollaboration.org/lib/project/126/.

Wilson-Kovacs, D., 2014. 'Backroom boys': Occupational dynamics in crime scene examination. *Sociology*, *48*(4), 763–779.

8

EMERGING FORENSIC GENETIC TECHNOLOGIES

Contested anticipations of legitimation, caution and social situatedness

Christopher James Lawless

Introduction

Recent years have seen significant scientific developments and advances in DNA technology regarded as presenting opportunities for forensic application. Technologies which promise faster ways of analysing genetic material, such as rapid DNA, and those which claim to generate significantly more investigative utility from individual DNA samples, such as forensic DNA phenotyping (FDP) and forensic genealogy, currently exist at something of a threshold moment. At time of writing, they have been developed and have found usage in some jurisdictions but have been received with a notable degree of epistemic and ethical caution elsewhere (see, for example, Toom et al 2016; Wienroth 2020). In some jurisdictions, concerns remain over the perceived need to validate such systems and whether their social and ethical impact is sufficiently well understood (Samuel and Prainsack 2018; Granja and Machado 2020; Granja et al 2020).

The emergence of these technologies is, however, indicative of the recognised speed at which biometric systems are developing (Scottish Parliament 2019). The complexity, rapidity of development and attendant increase in various biometric data forms and technologies test conceptual frameworks of innovation management and technological anticipation (Borup et al 2006). This literature has tended to focus on singular examples of technology (Korsnes 2016; Hielscher and Kivimaa 2019; Lawless 2021) rather than the divergent but potentially interlinked forms explored in this volume. Much sociology of expectations literature has also benefited from being able to examine singular empirical examples of technology over relatively long historical periods (Hielscher and Kivimaa 2019). Such an approach may yield ample historical data, but it means any analysis comes retrospectively through hindsight. How then to address complex technological forms whose

DOI: 10.4324/9780429322358-11

futures, due to their fast-developing nature, may already have at least partially arrived, in a world kept off balance by wider fast-moving events?

This chapter sees the partially realised nature of these technologies as an opportunity rather than a problem. It utilises this opportunity to explore interpretations of emerging forensic DNA technologies to illuminate how they render orderings of various domains fluid and malleable. Such domains include the natural and social sciences, ethics, law, commerce and society at large. The chapter draws on the conceptual framing of interpretive flexibility as developed via social construction of technology studies, which have highlighted the different perceptions of technology on the part of different stakeholders embedded in sociotechnical networks (Pinch and Bijker 1984). Differing standpoints and interpretations of technology may lead to contested plans and imaginaries (Lawless 2020). Framing technolegal anticipation – namely, anticipations of law-science-technology-society relations (Faulkner et al 2012) – in terms of embedded interpretive flexibility enables the chapter to illuminate differing normative interpretations of technology and sociotechnical engagement. Here, discursive responses to three notable examples of emerging forensic DNA methods are explored: rapid DNA analysis, forensic DNA phenotyping (FDP) and forensic genealogical testing. These three examples are presented in order to examine how their perceived status embodies differing assumptions concerning how science and wider society should engage, which, in turn, order relations between domains in different ways.

The chapter explores commentaries on these three technological examples as found in the literature. It elucidates a three-way typology of normative interpretations which reflects contrasting anticipations of forensic DNA technology. These commentaries were read by tracing various perceptions in their accounts of debates and issues which surround the three technological examples. These readings were also informed by discussions with stakeholders, including forensic scientists, social scientists, policy makers and regulators.

This account builds on critical public understanding of science perspectives which have been developed through social studies of forensics (or 'Forensic Studies') (Lynch 2009; Cole 2015; Lawless 2016). It suggests that a first series of normative interpretations is redolent of the so-called deficit model of public understanding of science, which assumes that lay audiences need to be educated and accept the transformative potential of science and technology (Bodmer 1985; Miller 2001). These are largely supportive of emerging forensic DNA technologies and perceive social benefits. A second series of normative interpretations relate to the 'surfeit' model, which posits that lay audiences assume too much scientific familiarity (Cole 2015). Here, it is advocated that these reflect epistemic caution on the part of scientific authority. The chapter seeks to advance forensic studies perspectives by considering a third variant of interpretations which have been previously termed 'social realist' to describe how extra-scientific concerns may be prioritised in sociotechnical engagements (Lawless 2016). These reflect deliberations around the perceived social risks of emerging technologies. It is suggested here that interpretations of social risk reflect a more balanced and interdependent ordering of scientific and social concerns which align with Jasanoff's (2004) notion of 'co-production'. This chapter proposes that, together, these three normative interpretations – deficit, surfeit and social realist – reflect different expectations of how science and technology are positioned relative to society at large. By exploring how these interpretations project differing notions of these technologies, this chapter argues that they represent contested anticipations and suggests that the varied science-society orderings which underpin them can be seen as distinct sociotechnical imaginaries: namely, competing 'collectively imagined forms of social life and social order' (Jasanoff and Kim 2009: 120).

Rapid DNA analysis

Introduction

Rapid DNA technology refers to methods which could replace the laboratory processing of forensic DNA profiles. Rapid DNA involves the installation of devices, approximately the size of desktop printers, in police custody suites from which DNA profiles can be collected from the biological material of arrestees (Turingan et al 2016; Murphy 2018). Such technology has been claimed to produce DNA profiles within a few hours, compared to three to five working days, which has been the norm previously in jurisdictions such as England and Wales.

Social benefits

A reading of discussions in the literature highlights a number of ways in which rapid DNA has been claimed to have a potentially transformative effect on policing practices and whose impact could extend even further. Some justifications for rapid DNA align with narratives of 'rationalization' and resource efficiency in policing, of science and technology being used unproblematically to speed up criminal justice processes (Steward 2016; Wilson-Wilde and Pitman 2017: 2; Morgan et al 2019). Rapid DNA has also been linked with the possibility of more timely decision-making about whether suspects should be held in custody or not. These supportive perceptions have gone further in claiming that rapid DNA could actually benefit civil liberties by exonerating individuals or eliminating them from enquiries much more quickly than existing processes (Steward 2016). Rapid DNA has furthermore been justified as a means that might reduce the number of items requiring analysis while improving and accelerating the process of identifying items that may yield useful evidence (Raymond 2015: 368). The possibility that rapid DNA could be linked to real-time comparison with DNA databases also forms part of these rationalisation narratives (Murphy 2018: 308).

One notable perceived resource efficiency concerns the possibility that rapid DNA may circumvent the need to send evidence to external laboratories, seen as costly and time consuming. Currently, however, there remains a requirement in some jurisdictions for qualified scientists to operate rapid DNA equipment (Murphy 2018: 309). Changes to the law may be required to allow for rapid DNA processing by other individuals such as police officers, yet this may be justified as circumventing the economic and time costs of relying on laboratory analysis (Murphy 2018). Other claimed possibilities include using rapid DNA to speed up identification checks at borders, together with suggestions that an individual's DNA could replace documented ID such as passports. Rapid DNA has also been claimed as beneficial for identifying persons who may have fallen victim to human trafficking (Steward 2016: 1150).

Rationalising commercial and technical drivers may lead rapid DNA to evolve rapidly in the future. It has been claimed that rapid DNA systems could subsequently process profiles from smaller quantities of DNA (Turingan et al 2016) or become more miniaturised, facilitating possible wider use in law enforcement contexts such as crime scenes (Steward 2016: 1162).

To summarise, discourses of justification for rapid DNA reflect a kind of technological determinism in promoting this technology as a driver of rationalisation and efficiency in policing. In doing so, such discourses also serve to potentially normalise and legitimise rapid DNA and the use of DNA as an identifier across a wider range of spaces by removing the perceived need for laboratory involvement.

Epistemic caution

Other interpretations, however, have adopted a more cautious view of rapid DNA. Rapid DNA systems have been seen by some forces, such as London's Metropolitan Police, as presenting numerous validation challenges relating to technical standardisation, user training, accreditation, and data storage and reporting before they have been deemed usable (Dolan 2016). Other agencies elsewhere, notably the US Federal Bureau of Investigation, have argued the need for rapid DNA to be subject to rigorous standard setting (Murphy 2018: 309). Some other commentaries have emphasised how rapid DNA systems use a more limited number of biological markers than laboratory-based methods and may be more suitable as an intelligence means of eliminating individuals from an investigation rather than as evidence to inculpate suspects (Mapes 2017). It has been questioned whether a reliance on DNA evidence via rapid DNA would have any significant impact on policing outcomes in terms of detections or convictions (Mapes et al 2016), which mirrors earlier critiques of police reliance on genetic material (McCartney 2006). Regardless of any resource efficiencies brought about by rapid DNA, interpretation of profile data has been identified as a concern (Gallop 2016). It has been argued that the rapid processing of profiles only assists with so-called source-level questions (Cook et al 1998), which refer to issues concerning the source of a DNA profile at a scene. Yet such questions do not fully address activity-level questions as to how a profile came to be deposited at a scene, for which police may require other case information (Gallop 2016). As with other emerging forensic technologies, it has been argued there could potentially be severe reputational costs for rapid DNA if it comes to be scrutinised in court and reliability and validity issues are identified (Dolan 2016). Rapid DNA processing has been claimed to be vulnerable to methodological issues such as difficulties in resolving mixed DNA profiles or contamination issues, and concerns have been expressed regarding how well informed police officers may be in such matters.

Concerns around the potential reputational costs of rapid DNA, the need to set standards and considerations of how DNA is used in investigations have been recognised in commentaries, but it remains to be seen how such questions may remain open to debate in the light of pressures to introduce rapid DNA as described earlier.

Social risks

Interpretations of social risk reflect a wider series of pressures, such as those on police forces in some jurisdictions to make budget cuts. Rapid DNA, however, has been regarded as presenting a risk to the forensic market in that it might reduce the amount of custom to commercial providers or force laboratories to downsize (Murphy 2018). Yet rapid DNA has also been claimed to present financial costs via the need for additional validation, accreditation and maintenance (Wilson-Wilde and Pitman 2017: 8–9).

The possibility of using rapid DNA in a wider context has led to ethical concerns being expressed about what the technology should be used for and raised questions about its possible lawfulness in jurisdictions such as the US, where it may violate the Fifth Amendment against self-incrimination if rapid DNA enables involuntary testing and possible matching of a suspect at a crime scene (Steward 2016: 1155–1158).

The introduction of rapid DNA into spaces such as custody suites or crime scenes could be said to intervene in the construction of boundaries over who constitutes 'experts' or 'non-experts' regarding the processing of DNA (Wilson-Wilde and Pitman 2017). Legislation could function as a shaper of 'expertise' (Steward 2016: 1162). Police officers who were previously not 'expert' scientists might gain their trappings if they gain the authority to operate rapid DNA. Questions have been raised over how far this 'expertise' might go and whether officers would be 'accredited' simply to load a sample and push buttons rather than to interpret profiles. Concerns have thus arisen over how these officers might perceive their own 'expertise' and who would bestow it (Wilson-Wilde and Pitman 2017)

To summarise, interpretations of social risk regarding rapid DNA reflect concerns about economic pressures on forensic laboratories. These relate to how they might change the nature and amount of work given to laboratories. In addition, the use of rapid DNA could be facilitated by having 'expertise' re-defined by legislators rather than scientists, which, in turn, raises ethical and operational issues.

Forensic DNA phenotyping

Introduction

DNA phenotyping broadly refers to methods which claim to infer a detectable genetically inherited characteristic (a 'phenotype') such as physical appearance or a medical condition from a DNA sample. Forms of forensic DNA phenotyping

(FDP) have been used in casework in a number of jurisdictions worldwide (Sankar 2010; Jong and M'charek 2018). Sankar (2010) described the first use of such methods in relation to the identification of Derrick Todd Lee from Louisiana in connection with a string of homicides reported during the late 1990s and early 2000s (Sankar 2010: 55). Early methods of discerning 'external visible characteristics' (EVCs) (Toom 2012) were also used to investigate the rape and murder of Marianne Vaatstra from Zwaagwesteinde in the Netherlands (Jong and M'charek 2018). DNA analysis in this case suggested the perpetrator was of Northern European appearance, a result which claimed to 'alleviate the social tensions surrounding the incrimination of asylum seekers' (Jong and M'charek 2018: 358). Vaatstra's body had been found close to a shelter for asylum seekers from the Middle East and North Africa (Sankar 2010: 56).

Social benefits

Parabon Nanolabs, a company which has received support from the US Department of Defence, has produced a method known as Snapshot, which it claims can produce facial images of persons from unknown DNA profiles. Snapshot has attracted much interest from law enforcement communities (Samuel and Prainsack 2018), and successes using FDP have been claimed (Murphy 2018). Supporters of FDP have framed it as a 'biological witness' (Kayser 2015: 45). These supporters claim it to be, at least in theory, a more accurate and epistemologically robust form of identification than human eyewitness testimony, which, it is claimed, might reduce racial discrimination. While supportive discourses concerning FDP are apparent in academic literature (Kayser 2015; MacLean and Lamperello 2014), there are instances of wider support for FDP elsewhere. One such example concerns responses to a case in Germany involving the rape and murder of a student in the university town of Freiburg. Around this time, there were calls for including FDP in German law. Lipphardt (2017) reported that media coverage of FDP subsequently became 'overtly positive', particularly in Freiburg itself (Lipphardt 2017). Support for FDP came from a variety of sources, including supporters of right-wing political groups, politicians, local police representatives and the media. Lipphardt reported that FDP supporters claimed the law needed updating to accommodate technological advances. Such discourses of justification for FDP reflect an interpretation of this technology as a sophisticated but unproblematic means of updating law enforcement procedures.

Epistemic caution

Elsewhere, however, other voices convey a different view of the scientific issues concerning FDP. Concerns have been expressed over whether investigators may feel pressured to use this new technology framed overly positively by media and politicians. Such related discourses may resemble the 'CSI effect', in which media depictions present a skewed view of forensic technology as authoritative and absolutely reliable (Lipphardt et al 2016; Jong and M'charek 2018). When Snapshot produced an image of a young African American male in connection with the murder of Candra Alston and her daughter in South Carolina, concerns were expressed over the accuracy and usefulness of the image (Lawless 2016) and the validity and reliability of the underlying science (Gannon 2017; Wienroth 2020; Samuel and Prainsack 2018). Other issues have been raised over regulatory standards, the level of legislation to which FDP may be subject and whether it might be vulnerable to critical courtroom scrutiny (Lawless 2016). Political pressure to allow the use of FDP in Germany was critiqued by academic communities, who claimed that complexities concerning the logical basis for interpreting FDP data had been underplayed (Lipphardt 2017). FDP has been associated with some negative high-profile casework outcomes. For example, Lipphardt et al (2016) describe how police investigating a series of murders in Heilbronn, Germany, in 2007 interpreted DNA evidence as wrongly incriminating members of the Sinti and Roma communities (Lipphardt et al 2016; Lipphardt 2017; Skinner 2018). To summarise, these interpretations thus caution against an overly simplistic or unproblematic view of FDP, instead emphasising scientific and operational complexities.

Social risks

Some commentaries have framed FDP as being embedded in a series of ethical concerns (Toom et al 2016: 2–3; Smith and Urbas 2011). The use of FDP to identify putative 'suspect' populations and the pressure on individuals identified with those populations to submit DNA to exonerate themselves have raised fears over the discriminatory targeting of certain populations and presumptions of innocence being challenged (Toom 2012; Toom et al 2016: 3–4). Much ethical discussion around FDP relates to concerns about how this technology may problematically construct conceptions of race. Skinner (2018) points to the implication of language (e.g., terms such as 'race', 'ethnicity' etc.) as reflecting a slippage between objective/subjective and social/biological accounts (Skinner 2018).

> Assertions also assume that markers of race and ethnicity are obvious and unproblematic and in the process gloss over questions of categorisation: why pick particular race categories, how should people be placed in or out of those categories, and what operational and social implications attach to these choices? . . . phenotype prediction does not, on its own, result in a single felon but creates a group of suspects.
>
> *(Skinner 2018: 12)*

Other ethical issues relate to

> the right of people not to know what their DNA tells about propensities for diseases or other propensities, data protection and privacy . . . the risk of

stigmatization and discrimination, and the vision of a slippery slope leading, ultimately, to eugenics.

(Koops and Schellekens 2008: 160; see also Perepechina 2013)

Here, these perceptions raise concerns about forensic scientists constructing categorisations which may be invalid and discriminatory, holding adverse social consequences.

Another factor which has been identified as a potential constraint on FDP is cost. It has been claimed that realising the scientific potential of FDP would require 'large collaborative efforts' (Kayser 2015: 34) with commensurately high levels of research funding needed (ibid: 46). The role of private enterprise in developing FDP methodology and its use has raised questions about how it may be scrutinised and regulated and the potential to cite commercial confidentiality to withhold validation data (Wienroth 2020). This has previously been employed as a response to scientific criticism in other examples of emerging forensic technology (Lawless 2013). In summary, then, a wider series of issues relate to the collective negotiation of ethical and scientific boundaries, concerns about co-production of ideas about race and other categories; regulatory gaps; and the influence of commercial imperatives.

Forensic genealogy

Introduction

Commercial services which offer genealogical testing via DNA analysis have become popular with the public in recent years, but such services have also been exploited for the purposes of criminal investigation (O'Leary 2018; Kennet 2019; Phillips; Ram et al 2019; Wickenheiser 2019). Unsuccessful attempts in earlier investigations (Kennet 2019) did not impede police enthusiasm in some jurisdictions, and genealogical analysis has since come to be associated with some high-profile case outcomes. In 2015, it was reported that genealogical analysis produced a lead which prompted the arrest of Bryan Patrick Miller for Arizona's Canal Killer murders (Ram et al 2019; Phillips 2018). Another longstanding unsolved murder, known as the Buckskin Girl case (referring to a murder victim's jacket) was also pursued via genealogical analysis. GEDMatch, a public genealogy service, returned an apparent familial link (Kennet 2019: 106).

One of the most prominent forensic interventions into genealogy concerns the Golden State Killer case (O'Leary 2018; Kennet 2019; Ram et al 2019; Wickenheiser 2019). The Golden State Killer was linked to a string of rapes, murders and burglaries committed in California between 1976 and 1986. A DNA profile of the killer had been obtained from a rape kit. The profile was uploaded to GEDMatch. which identified between 10 and 20 distant relatives. Police collaborated with genealogist Barbara Rae-Venter to produce a family tree, which identified Joseph James DeAngelo as the main suspect. In 2018, a DNA sample was covertly obtained from the door of DeAngelo's car and another from a tissue in a refuse bin. Both these DNA profiles matched crime scene samples linked to the Golden State Killer. Rae-Venter has subsequently assisted with other investigations (Kennet 2019).

Social benefits

Commercial genetic genealogy services have drawn widespread public interest. Regalado (2018) claims 2017 to have been a landmark year in uptake, in which over 12 million individuals in the US analysed their DNA via these services, a rise of over double the previous year, amid claims that 1 in 25 US citizens had access to personal genetic data (Regalado 2018). This figure reportedly rose to 26 million in 2018 (Regalado 2019). Apparent successes such as the Golden State Killer case were accompanied by positive news coverage and evidence of supportive public attitudes towards the forensic use of genealogical databases, at least for violent crimes (O'Leary 2018; Kennet 2019, Greytak et al 2019). It was claimed that more individuals registered with GEDMatch following the Golden State Killer's arrest (Greytak et al 2019: 107). One public opinion survey conducted in the USA by Guerrini et al (2018) suggested support for police searches of commercial genealogy resources, for companies to disclose customer information to police and for police to submit fake profiles or use false names (Guerrini et al 2018). This survey indicated notable support for these practices to investigate violent crime and crimes against children and missing persons but less support when in connection with nonviolent cases.

Greytak et al (2019) discussed the ways in which public users are educated about the status of genetic material submitted to commercial genealogy services. They claim that GEDMatch made customers sufficiently aware that their data might be accessed by police prior to analysis (Greytak et al 2019: 106–107). TV advertising has urged the public to supply their DNA to genealogical databases to assist criminal investigations (Kennet 2019: 112). Such a call has a potential global reach. New Zealand citizens were encouraged to submit data to investigate crimes elsewhere (Kennet 2019: 114). A US congresswoman appealed to the provider 23andMe to help reunite Mexican families (Syndercombe Court 2018: 204). It has been reported that UK police forces have expressed interest in the use of genealogical testing (Biometrics and Forensics Ethics Group 2018). Attempts to justify uploading crime scene samples to genealogical databases and surreptitious sampling to solve violent crimes, along with other perceived utilities, indicate the perception of the scientific power of genealogy on the part of some police, the public and policy makers.

Epistemic caution

Concerns have been raised about police understanding of the science of genealogical testing and whether it might lead investigators down misleading routes (Wickenheiser 2019: 119; Scudder et al 2019). Such issues relate to the levels of accreditation among genealogists and the capacity of police to guarantee the competency of collaborators (Kennet 2019: 113). Like FDP, genealogy is regarded by many as only generating leads. Conventional DNA testing from an arrestee, whether surreptitiously or consciously, is still necessary to link a suspect with crimes.

Other practicalities have been perceived as presenting a challenge to the application of genealogy. These include perceived issues associated with conventional DNA analysis, such as problems involving the resolution of DNA profile mixtures, contamination and partial matches. It has been claimed that certain individuals may have to provide elimination samples (Wickenheiser 2019: 121), which some labs do not provide for. It has been noted that investigators using genealogical testing have to get samples tested via single nucleotide polymorphism (SNP) analysis, but only a small number of providers offer this service (Kennet 2019: 109), and it has been claimed that most US labs are only equipped to work with short tandem repeats (STRs) (Wickenheiser 2019: 119). In the UK, the Forensic Science Regulator has raised concerns about scientific standards (Biometrics and Forensics Ethics Group 2018), and it has been claimed that many commercial methodologies have not been validated for forensic use (Kennet 2019: 109), which may necessitate consideration of many processes (Wickenheiser 2019: 121–122). Such voices assert that investigative processes are time consuming and act as a limiting factor influencing what cases they might be used for. They point out, for example, that the Golden State Killer investigation took four months (Wickenheiser 2019: 115). These issues suggest another discourse whereby it is assumed that non-scientific users might adopt an overly simplified view of this science.

Social risks

A host of even wider social issues have been voiced concerning the forensic use of DNA genealogy, including ethical concerns, admissibility issues and matters of user protection. From an ethical point of view, one claimed issue concerns exceptionality: namely, questions of how to determine or justify which cases should be investigated via genealogical methods (Granja and Machado 2019). It has been stated that US public attitudes differ regarding support for use in serious cases versus non-serious cases (Kennet 2019) in the context of changing definitions of 'violent crime' (Ram and Roberts 2019).

The Golden State Killer case raised ethical concerns in some quarters over the surreptitious sampling of suspect DNA, together with concerns about possible function creep regarding the use of commercial databases for purposes for which they had not been anticipated. While some surveys suggest high levels of public support, other commentaries claim some people are uncomfortable with commercial genealogical databases being used for criminal investigations (O'Leary 2018). The popularity of commercial DNA genealogy testing and its portrayal in advertising reflect concerns that such imagery reinforces idea that kinship identity is purely genetically determined rather than also socially shaped (Haimes 2006). Other concerns relate to the possibility that a reliance on genealogical testing may reinforce anachronistic ideas about criminality running in families (Syndercombe Court 2018: 204) or that forensic genealogy might lead toward a kind of reductionist-determinist framing of race/ethnicity and identity in general (Roth and Ivemark 2018). Further expressed issues relate to whether the application of genealogical methods could potentially disrupt family identity (Bowman and Grindrod 2019), whether genealogy testing should be combined with social media searching and whether surreptitious testing is ever acceptable. The accuracy of ancestry and ethnicity claims have been challenged (Wickenheiser 2019).

Ethical concerns extend to legal and commercial matters. For example, in one case, the provider FamilyTreeDNA allegedly allowed law enforcement to access data without telling customers (Ram and Roberts 2019). Commentaries have drawn attention to the existence of legal permissions allowing police to access third-party data (Ram et al 2019). Other commentaries have drawn attention to differences in the law which exist across US states (Wickenheiser 2019) and between different parts of the world. For example, due to data protection laws, EU-based users are allowed to opt out of police investigations, in contrast to the US (Ram and Roberts 2019).

To summarise, these discourses express concern about the extent to which genealogy might enable law enforcement to intervene or interfere in people's lives and the ways in which genealogy might shape or reinforce the public's assumptions about family identity.

Discussion

This chapter has interrogated three prominent examples of emerging forensic DNA technologies. In doing so, this account has elucidated a series of normative interpretations relating to them. These normative interpretations reflect distinct sets of expectations and concerns on the part of stakeholders, but more significantly, they represent various projected configurations of relations between science and society at large. We can see through these examples how forensic technology may be discursively positioned hegemonically and influentially over society as a more strictly bounded domain of scientific concern or framed in an interdependent but potentially contentious relationship with society. In this final section, I discuss the further implications of these different framings.

The first such mode of interpretation emphasises social benefits, in a form which projects a technologically determinist technolegal world. This frames technologies as unproblematically beneficial and socially transformative. While some scientific voices have promoted technologies in this way, this interpretation is notably reflected in public, media, political and police support for rapid DNA, FDP and genealogical analysis. This interpretation frames these technologies as improving and updating law enforcement through the increased rationalisation they seemingly bring. The perceived trust in science and technology underpins assertions that these technologies should be used widely. The claimed 'rationalizing' benefits of technology are emphasised, while ethical imperatives feature less prominently, other than the importance of allowing technology to support justice and maintain public safety. In the example of FDP, this technology was promoted as superior to eyewitness testimony. This interpretation is possibly also reflected in public support for the use of genealogical analysis in law enforcement.

A second normative interpretation identified here as associated with epistemic caution could be said to be a sceptical mode, thus projecting a more conditional and cautious technolegal world. This invokes the observations of some authors who have drawn attention to concerns that media depictions may lead lay audiences to assume too much knowledge of forensic science, rendering them oblivious to scientific complexities and practicalities (Cole 2015; Lynch 2009). This sceptical mode frames forensic technology as epistemically conditional and prioritises rigorous scientific oversight and standards through expert consensus. This interpretation emphasises the potential reputational risks if scientific shortcomings are overlooked (Lawless 2016). In contrast with interpretations of social benefit, more cautious voices emphasise the necessity for expert and regulatory scrutiny. These interpretations also emphasise the limits to the information such technology can provide in the course of criminal investigations. Such sceptical interpretations may be expressed by those who have a direct concern with scientific standards and arguably serve to delineate a narrower cohort claiming certain forms of scientific expertise.

The third normative interpretation identified in this chapter associated with social risk frames emerging forensic technologies as interdependently embedded in society and projects a co-productionist technolegal world (Jasanoff 2004). It interprets these technologies in terms of social impact, perceiving technological possibilities to be co-constructed with extra-scientific domains such as law, commerce, jurisdictionality, geography, ethics etc. Here, social risks and possibilities intertwine with technological development. These include:

1. The risky co-production of expertise (as in laws potentially being changed to allow police officers to become 'expert' operators of rapid DNA systems rather than accredited scientists)
2. The questionable use of certain racial or ethnic categories to classify phenotypic data and the potential blurring of social and biological boundaries in the case of FDP or their ethically problematic conflation in the case of genealogical analysis.
3. The possibility of constructing differential subjectivities, or biometric winners and losers, as in the example of EU data protection laws preventing genealogical analysis of its citizens while others may be subject to it, governance gaps leading to the use of certain technologies in some US states rather than others and, as before, the contentious construction of 'suspect' populations
5. The economic risks of developing forensic technologies which may lead to high sunk costs and the ensuing 'lock-in' of technologies which may be suboptimal and ethically problematic
6. The potential barriers to scientific scrutiny and the ensuing risks of emerging technologies through claimed extra-scientific factors such as commercial confidentiality.

These normative interpretations represent a set of discursive spaces where a series of domains – scientific, policing, legal, economic, ethical, social etc. – intermingle. Projected presents and futures of the three examples of emerging forensic DNA methods reflect and mobilise differing orderings of these domains and assumptions concerning how science and wider audiences should engage. Here, social benefits may stress policing responsibilities and rationalisation and epistemic caution gives primacy to scientific actors while social risks frame a problematically interdependent relationship between forensic technology and a host of projected social and ethical factors. This latter notion of co-production which manifests itself via social risk is only one possible framing of forensic DNA technologies. It intermingles with the technological determinism of social benefits and the scientific hegemony of epistemic caution. Co-production may thus just be one projected ontological and epistemological possibility, an interpretation which may not be shared by all stakeholders and which might largely remain limited to the purview of social science. This raises a provocative question: Might co-production itself be a socio-technical imaginary promoted by social scientists, possibly partially shared by some stakeholders but not by others, competing with other assumptions and visions for scientific and social engagement?

This does not, however, lead to a regressive dead end. The interpretive flexibility of emerging DNA methods opens up other potentially fruitful questions: namely, how actors might anticipate the way other stakeholders imagine technolegal futures. This suggests new ways of thinking in terms of conceiving how actors understand their own perceptions and anticipate those of others, suggesting new framings of technolegal worlds as sites of negotiation and co-ordination. Understanding how these assumptions and visions contend to shape emerging forensic technologies may significantly help capture technolegal world building in progress.

Bibliography

Biometrics and Forensics Ethics Group (2018) 'Meeting minutes of 20 September 2018', https://assets.publishing.service.gov.uk/government/uploads/system/uploads/attachment_data/file/837706/BFEG_minutes_-_20_September_2018.pdf Accessed 1 September 2020.

Bodmer, W. (1985) *The Public Understanding of Science*. London: Royal Society.

Borup, M., Brown, N., Konrad, K and Van Lente, H (2006) The sociology of expectations in science and technology. *Technology Analysis & Strategic Management* 18(3–4): 285–298.

Bowman, C.E. and Grindrod, P. (2019) 'Kinship, familial searching and biometrics', June 2019. www.researchgate.net/publication/333982366_Kinship_Familial_Searching_and_Biometrics Accessed 31 August 2020.

Cole, S. A. (2015) 'A surfeit of science: The "CSI effect" and the media appropriation of 23 the public understanding of science', *Public Understanding of Science*, 24(2): 130–146.

Cook, R., Evett, I., Jackson, G., Jones, P. J. and Lambert, P. (1998a) 'A model for case 34 assessment and interpretation', *Science and Justice*, 38(3): 151–156.

Dolan, A. (2016) 'The use of new and emergent forensic genetic technologies', ESRC Research Seminar series on genetics, technology, security and justice. Crossing, contesting and comparing boundaries Seminar 2: Comparing stakeholder discourses about genetic technologies. Durham University 16 March 2016.

Faulkner, A., Lange, B. and Lawless, C. (2012) 'Introduction: Material worlds: Intersections of law, science, technology and society', *Journal of Law and Society*, 39(1): 1–19.

Gallop, A. (2016) 'Upsides and downsides associated with the introduction of new DNA technologies into operational forensic science', ESRC Research Seminar series on genetics, technology, security and justice. Crossing, contesting and comparing boundaries seminar 2: Comparing stakeholder discourses about genetic technologies. Durham University 16 March 2016.

Gannon, M. (2017) 'Amazing DNA tool gives cops a new way to crack cold cases: DNA phenotyping can produce a sketch of the suspect. But is it ready for primetime?' *NBC-News*. https:// www.nbcnews.com/mach/science/amazing-dna-tool-gives-cops-new-way-crack-coldcases-ncna781946. Accessed 28 September 2022.

Granja, R. and Machado, H. (2019) 'Ethical controversies of familial searching: The views of stakeholders in the United Kingdom and in Poland', *Science, Technology & Human Values*, 44(6): 1068–1092.

Granja, R. and Machado, H. (2020) 'Forensic DNA phenotyping and its politics of legitimation and contestation: Views of forensic geneticists in Europe', *Social Studies of Science*. First published 30 July 2020. https://doi.org/10.1177/0306312720945033.

Granja, R., Machado, H. and Queiros, F. (2020) 'The (de)materialization of criminal bodies in forensic DNA phenotyping', *Body & Society*, 27(1): 60–84.

Greytak, E. M., Moore, C. and Armentrout, S. L. (2019) 'Genetic genealogy for cold case and active investigations', *Forensic Science International*, 299: 103–113.

Guerrini, C. J., Robinson, J. O., Petersen, D. and McGuire, A. L. (2018) 'Should police have access to genetic genealogy databases? Capturing the golden state killer and other criminals using a controversial new forensic technique', *PLOS Biology*, 16(10): e2006906.

Haimes, E. (2006) 'Social and ethical issues in the use of familial searching in forensic 19
investigations: Insights from family and kinship studies', *Journal of Law, Medicine and* 20
Ethics, 34(2): 263–276.

Hielscher, S. and Kivimaa, P. (2019) 'Governance through expectations: Examining the long-term policy relevance of smart meters in the United Kingdom', *Futures*, 109: 153–169.

Jasanoff, S. (2004) *States of Knowledge: The Co-Production of Science and the Social Order*. London: Routledge.

Jasanoff, S. and Kim, S.-H. (2009) 'Containing the atom: Sociotechnical imaginaries and nuclear power in the United States and South Korea', *Minerva* 47(2): 119–146.

Jong, L. and M'charek, A. (2018) 'The high-profile case as 'fire object': Following the Marianne Vaatstra murder case through the media', *Crime Media Culture*, 14(3): 347–363.

Kayser, M. (2015) 'Forensic DNA phenotyping: Predicting human appearance from crime scene material for investigative purposes', *Forensic Science International*, 18: 33–48.

Kennet, D. (2019) 'Using genetic genealogy databases in missing persons cases and to develop suspect leads in violent crimes', *Forensic Science International* 301: 107–117.

Koops, B-J. and Schellekens, M. (2008) 'Forensic DNA phenotyping: Regulatory issues', *Columbia Science & Technology Law Review*, 9: 158–202.

Korsnes, M. (2016) 'Ambition and ambiguity: Expectations and imaginaries developing offshore wind in China', *Technological Forecasting and Social Change*, 107: 50–58.

Lawless, C. J. (2013) 'The low-template DNA profiling controversy: Biolegality and 6
boundary work among forensic scientists', *Social Studies of Science*, 43(2): 191–214.

Lawless, C. J. (2016) *Forensic Science: A Sociological Introduction*. London and New York: Routledge.

Lawless, C. J. (2020) 'Assembling airspace: The Single European Sky and contested transnationalities of European air traffic management', *Social Studies of Science*, 50(4): 680–704.

Lawless, C. J. (2021) 'The evolution, devolution and distribution of UK biometric imaginaries', *BioSocieties*. First published 5 May 2021. https://doi.org/10.1057/s41292-021-00231-x.

Lipphardt, A. (2020) 'The invention of the phantom of Heilbronn', *Journal for European Ethnology and Cultural Analysis*, 4(1): 49–69.

Lipphardt, V. (2017) 'Adapting the law to the technology? Germany's current debate on DNA phenotyping', ESRC Seminar Series on Genetics, Security and Justice: Crossing, Contesting and Comparing Boundaries. Newcastle-upon-Tyne 24 March 2017. www.ncl.ac.uk/media/wwwnclacuk/policyethicsandlifesciences/files/Lipphardt_slides.pdf. Accessed 31 August 2020.

Lipphardt, V., Lipphardt, A., Buchanan, N., Surdu, M., Toom, V., Wienroth, M., Mupepele, A. C., Cedric Bradbury, C. and Lemke, T. (2016) 'Open Letter on critical approaches to forensic DNA phenotyping (FDP) and bio-geographical ancestry (BGA)', Published online 8 December 2016.

Lynch, M. (2009) 'Science as a vacation: Deficits, surfeits, PUSS, and doing your own job', *Organization*, 16(1): 101–119.

Machado, H. and Granja, R. (2018) 'Ethics in transnational forensic DNA data exchange in the EU: Constructing boundaries and managing controversies', *Science as Culture*, 27(2): 242–264.

MacLean, C. E. and Lamperello, A. (2014) 'Forensic DNA phenotyping in criminal investigations and criminal courts: Assessing and mitigating the dilemmas inherent in the science', *Recent Advances in DNA and Gene Sequences*, 8(2): 104–112.

Mapes, A. A. (2017) 'Rapid DNA technologies at the crime scene: "CSI" fiction matching reality', PhD Thesis Chapter 10, University of Amsterdam. https://pure.uva.nl/ws/files/19345931/Chapter_10.pdf Accessed 1 November 2019.

Mapes, A. A., Kloosterman, A. D., de Poot, C. J. and van Marion, V. (2016) 'Objective data on DNA success rates can aid the selection process of crime samples for analysis by rapid mobile DNA technologies', *Forensic Science International*, 264: 28–33.

McCartney, C. (2006) 'The DNA expansion programme and criminal investigation', *British Journal of Criminology*, 46(2): 175–192.

Miller, S. (2001) 'Public understanding of science at the crossroads', *Public Understanding of Science*, 10: 115–120.

Morgan, R., Illidge, S. and Wilson-Wilde, L. (2019) 'Assessment of the potential investigative value of a decentralised rapid DNA workflow for reference DNA samples', *Forensic Science International*, 294: 140–149.

Murphy, E. (2018) 'Forensic DNA typing', *Annual Review of Criminology*, 1: 497–515.

O'Leary, D. E. (2018) 'DNA mining and genealogical information systems: Not just for finding family ethnicity', *Intelligent Systems in Accounting, Finance and Management*, 25: 190–196.

Perepechina, I. O. (2013) 'Legislative framework and value of the forensic DNA examination of health-related information for crime investigation', *Forensic Science International Genetics Supplement Series* 4: e360–e361.

Phillips, C. (2018) 'The golden state killer investigation and the nascent field of forensic genealogy', *Forensic Science International*, 36: 186–188.

Pinch, T. J. and Bijker, W. E. (1984) 'The social construction of facts and artefacts: Or how the sociology of science and the sociology of technology might benefit each other', *Social Studies of Science* 14(3): 399–441.

Ram, N., Guerrini, C. J. and McGuire, A. L. (2019) 'Genealogy databases and the future of criminal investigation', *Science*, 360(96393): 1078–10791.

Ram, N. and Roberts, J. L. (2019) 'Forensic genealogy and the power of defaults', *Nature Biotechnology*, 37: 707–708.

Raymond, T. (2015) 'The musings of a forensic scientist retiree', *Australian Journal of Forensic Sciences*, 47(4): 365–370.

Regalado, A. (2018) '2017 was the year consumer DNA testing blew up', *MIT Technology Review*, 12 February 2018. www.technologyreview.com/2018/02/12/145676/2017-was-the-year-consumer-dna-testing-blew-up/ Accessed 23 August 2020.

Regalado, A. (2019) 'More than 26 million people have taken an at-home ancestry test', *MIT Technology Review*, 2 November 2019 www.technologyreview.com/2019/02/11/103446/more-than-26-million-people-have-taken-an-at-home-ancestry-test/ Accessed 23 August 2020.

Roth, W. D. and Ivemark, B. (2018) 'Genetic options: The impact of genetic ancestry testing on consumers' racial and ethnic identities', *American Journal of Sociology*, 124(1): 150–184.

Samuel, G. and Prainsack, B. (2018) 'Forensic DNA phenotyping in Europe: Views "on the ground" from those who have a professional stake in the technology', *New Genetics & Society*, 38(2): 119–141.

Sankar, P. (2010) 'Forensic DNA phenotyping: Reinforcing race in law enforcement', In *What's the Use of Race?* eds. I. Whitmarsh and D. S. Jones, 49–62. Cambridge, MA: MIT Press.

Scottish Parliament (2019) *Scottish Biometrics Commissioner Bill Stage 1 Report*. Edinburgh: Scottish Parliament.

Scudder, N., Robertson, J., Kelty, S. F., Walsh, S. J. and McNevin, D. (2019) 'A law enforcement intelligence framework for use in predictive DNA phenotyping', *Australian Journal of Forensic Sciences*, 51(Sup 1), S255–S258.

Skinner, D. (2018) 'Forensic genetics and the prediction of race: What is the problem?' *BioSocieties*. https://link.springer.com/article/10.1057/s41292-020-00193-6 Accessed 31 August 2020.

Smith, M. and Urbas, G. (2011) 'Regulating new forms of forensic DNA profiling under Australian legislation: Familial matching and DNA phenotyping', Australasian Institute of Judicial Administration Conference, Sydney 7–9 September 2011.

Steward, E. R. (2016) 'Discussion and evaluation: The legality and use of rapid DNA technologies', *University of Missouri-Kansas City Law Review*, 84(4): 1134–1162.

Syndercombe Court, D. (2018) 'Forensic genealogy: Some serious concerns, *Forensic Science International: Genetics*, 36: 203–204.

Toom, V. (2012) 'Bodies of science and law: Forensic DNA profiling, biological bodies, 6 and biopower', *Journal of Law and Society*, 39(1): 150–166.

Toom, V., Wienroth, M., M'charek, A., Prainsack, B., Williams, R., Duster, T., Heinemann, T., Kruse, C., Machado, H. and Murphy, E. (2016) 'Approaching ethical, legal and social issues of emerging forensic DNA phenotyping (FDP) technologies comprehensively: Reply to "Forensic DNA phenotyping: Predicting human appearance from crime scene material for investigative purposes" by Manfred Kayser', *Forensic Science International: Genetics*, 22: E1–E4.

Turingan, R.S., Vasantgadkar, S., Palombo, L., Hogan, C., Jiang, H., Tan, E. and Selden, R.F. (2016) 'Rapid DNA analysis for automated processing and interpretation of low DNA content samples', *Investigative Genetics*, 7(2). eCollection.

Wickenheiser, R. A. (2019) 'Forensic genealogy, bioethics and the Golden State Killer case', *Forensic Science International Synergy* 1: 114–125.

Wienroth, M. (2018) 'Governing anticipatory technology practices. Forensic DNA phenotyping and the forensic genetics community in Europe', *New Genetics & Society*, 37(2): 137–152.

Wienroth, M. (2020) 'Socio-technical disagreements as ethical fora: Parabon NanoLab's forensic DNA Snapshot™ service at the intersection of discourses around robust science, technology validation, and commerce', *BioSocieties*, 15(1): 28–45.

Wilson-Wilde, L. and Pitman, F. (2017) 'Legislative and policy implications for the use of rapid DNA technology in the Australian context', *Forensic Science Policy & Management*, 8(1–2): 1–12.

PART III
Issues of legitimacy

9

SYSTEMIC (MIS)TRUST IN TECHNOLEGAL WORLDS

Three key trust relationships in forensic genetics

Matthias Wienroth

Introduction

In their 2019 report *Forensic science and the criminal justice system: a blueprint for change*, the United Kingdom (UK) House of Lords Science and Technology Select Committee dedicates an entire chapter to 'Ensuring trust in forensic science', arguing that:

> For forensic science to contribute effectively to the criminal justice system the science must be trustworthy. Two key components of this are quality standards and training.
>
> *(House of Lords Science and Technology Select Committee, 2019: 22, §74)*

With a focus on accreditation of forensic service providers and the training of its practitioners, that chapter reflects concerns in the UK forensic community about the scientific basis for forensic applications and the regulation of their use in practice. While not explicitly mentioned, other sources for trust in the criminal justice system's uses of forensic sciences and technologies are also suggested by the authors throughout the report. The report identifies a lack of high-level leadership and a unified strategic approach to forensic science, consistency in procurement and funding for forensic services, responsiveness to the need for new services, fair access to forensic testing and funding for innovative research.

> Unless these failings are recognised and changes made, public trust in forensic science evidence will continue to be lost and confidence in the justice system will be threatened. Crimes may go unsolved and the number of miscarriages of justice may increase. Furthermore, world-leading specialist expertise will

DOI: 10.4324/9780429322358-13

> be under-used, and England and Wales may never regain its reputation as holding the international benchmark for forensic science.
>
> *(Ibid.: 3–4)*

In the context of recent failures of the UK's forensic science market to provide a stabilised and diversified forensic service provision, aggravated by the 2012 closure of the government-owned company Forensic Science Service and high-profile forensic blunders and miscarriages of justice, the report presents a stark warning that the entire criminal justice system may suffer from irreparable loss of trust (or confidence, as the report puts it) should the recommendations in their 'blueprint for change' be ignored.

In the UK and elsewhere, trust is a central element in the discourse around public perceptions of science (Engdahl and Lidskog, 2014; Haerlin and Parr, 1999; Wynne, 2006). Trust is often portrayed as a requirement for the proper social organisation of sciences and technologies, especially when it comes to questions of uncertainty and of risk (Jacob and Hellström, 2000; Lidskog, 1996). The recurrent and powerful preoccupation with trust in relation to the use of DNA in the criminal justice system suggests that DNA evidence has simultaneously acquired 'an unprecedented degree of trust' and distilled 'a strong sense of fear' in stakeholder groups, based on:

> An allegedly unlimited evidential power, along with an also allegedly high level of sophistication and complexity which is commonly thought to be inaccessible to non-experts. DNA has become a 'black box' in the double meaning of the phrase: it has assumed the role of the ultimate and indestructible recorder of our 'secret' genetic code, and the role as an impenetrable and incomprehensible truth machine.
>
> *(Amorim, 2012: 259)*

Such contests – between risk, confidence and fear, and between credibility and implausibility – show that the deployment of science in support of criminal justice objectives is subject to particular trust relationships.

As I show in this contribution on the example of forensic genetics, some features of (especially adversarial) criminal justice systems treat trust as a local and fragile achievement, subject to constant and rigorous testing. This treatment is clearly visible in the use of methods for interrogating actors' claims to trustworthiness and credibility, methods that are rarely authorised in other social settings. This restlessness about the fragility of the achievement of trust is a constitutive and distinctive feature of criminal justice uses of science. While in the following analysis and discussion, I draw from examples often specific to the UK context and adversarial criminal justice systems, the chapter's findings about trust relationships offer general insights into the ways in which trust in forensic genetics is negotiated. I explore some of the uses of the term *trust* in context, together with the types of people who are invoked as part of forensic trust relationships in order to explore the role of trust in organising technolegal worlds around the use of genetics in the criminal justice system.

Forensic genetics and trust

The story of the rapid rise of DNA profiling as a forensic tool is widely known and often told (Jasanoff, 1995; Lynch et al., 2008; M'charek et al., 2013; Prainsack, 2010; Williams and Johnson, 2008). From its early uses, the question of trust in the technology has always been central to its challenge or acceptance and, thus, its potential utility to the criminal justice system. The early history of the use of forensic DNA evidences the strategic nature of claim making about the reliability of science and technology, as well as the collection and use of DNA samples by the police and deployment of forensic evidence in trials. A long process of negotiating changes in extraction, chain of custody, analysis and visualisation methods following several courtroom controversies in the USA and the UK certainly contributed to the increasing support by judicial and scientific authorities of the general acceptance of DNA profiling and matching, thus creating technolegal worlds in which DNA profiling has become trusted as the 'gold standard' of forensic evidence (Jasanoff, 1995; Kaye, 2010; Lynch et al., 2008). While many technological and organisational innovations which enable further and wider uses of DNA analysis are celebrated, there have also been critical voices warning of the dangers of 'function creep', of inadequate regulation, and of the risk to 'trust' when such innovations are introduced and expanded without sufficiently inclusive deliberation.

The UK was the first country to set up a central national forensic DNA database (NDNAD) in 1995. From 2000 onwards, a concerted effort was made by the then-UK government to expand the collection, profiling and databasing of DNA, accompanied and enabled by significant legislative changes. As a result, the NDNAD grew quickly to contain profiles of the equivalent of 10% of the entire population by 2010. While public trust debates related to concerns about proportionality, privacy, dignity, equality and lack of transparent governance have been an integral part of the debate in the UK since the mid-2000s (e.g., Human Genetics Commission, 2009; GeneWatch UK, 2010, 2011; Metropolitan Police Authority Civil Liberties Panel, 2011; Roberts, 2013; Home Office, 2013), in this contribution, I focus on stakeholders within science, policing and the courts in order to explore the technolegal worlds created in the core domains of forensic genetics practice in the criminal justice system. For a comprehensive analysis of the history of DNA databasing in the UK, see Skinner and Wienroth (2019).

Systemic trust

There are many approaches to understanding and defining trust and its terms (e.g., Barbalet, 2006; Frederiksen, 2012; Khodyakov, 2007; Möllering, 2001). Trust is 'a matter of judgement and action, in conditions of less than perfect information' (O'Neill, 2004: 271) and, consequently, an expression of expectations about the future (Luhmann, 1979: 25). Participants in a working trust relationship can rely on, for example, expert knowledge to provide solutions to complex questions without needing to become experts in the different fields that contribute to providing such solutions to shared problems. In this way, trust relationships enable the division of labour while permitting risk calculations and capacity to endure uncertainty. At the micro level, trust exists between individuals; at the macro level, such individual relationships become part of sets of practices and institutions that make up a system. Forensic science is one such set of practices and institutions; policing and the courts are others, constituting what we know as the criminal justice system (from here: CJS). Trust-related 'problems are posed in terms of the maintenance of stability of action systems' (ibid: 5). For the CJS, the provision of security, justice and fairness raises some of the complex questions that need to be addressed by answering questions about culpability, evidence, detection and so forth. Trust relationships are articulated as core means of producing stability in (and of) the CJS. The use of forensic technologies in policing to provide intelligence in the process of detection and evidence for trials, among other things, represents one trust relationship. Forensic work helps in reducing complexity for stakeholders by shifting the burden of the scientific and operational knowledge required to make informed decisions from investigators and judges to forensic practitioners and expert witnesses. In doing so, the CJS brings together law and science, in themselves systems of practices, institutions and trust relationships. At this point, trust between individuals is largely overlaid by what I call *systemic trust*.

Systemic trust is a functional element, acting as 'social glue' by creating trust relationships between practices and institutions, such as professional groups. This glue holds together domains that otherwise follow their own rules: here, specifically, science and the law, forensic service provision and prosecution/police investigation. Arguably, trust helps create communities of practice by providing 'stickiness through being able to bring together the main players in the field' (Molyneux-Hodgson and Meyer, 2009: 141). Expanding the metaphor, systemic trust also acts as 'procedural oil' for a system's running by providing legitimacy (see also Hough, 2021: 13–30) to certain practices – such as DNA profiling and databasing – and, as a result, to a system's stability.

In this chapter, I do not offer a semantic analysis of trust concepts – I refer to trust, credibility, confidence and trustworthiness as key concepts of trust relationships without developing a nomenclature. Instead, I scrutinize three core trust relationships that stakeholders articulate vis-à-vis forensic genetics in the CJS.

Methodology

The data used in this chapter were derived from 16 semi-structured interviews conducted by the author between 2012 and 2016, from transcripts made available to the author from four multi-stakeholder meetings held between 2004 and 2010 and from a variety of documents. Interview respondents were academic forensic geneticists, forensic practitioners, barristers and police officers in the UK, together with forensic scientists and practitioners across Europe. Forensic stakeholders involved in the records of meetings included academic scientists, barristers and police officers, and these meetings took place in Dublin (2004), Durham (2004), London (2010) and Oxford (2010). Public documents include policy reports, academic publications, press releases and publications by third-sector organisations. The data refer especially to the CJS in England and Wales, but the inclusion of individuals from other jurisdictions in interviews and meetings makes it possible to offer some wider observations. The data from these sources have been analysed using qualitative thematic analysis along relevant lines of trust and its synonyms and antonyms.

Narrating, enacting and challenging trust in forensic genetics

When trust is called for by forensic genetics stakeholders, invoking the need for or loss of trust, two things happen simultaneously: trust is related to a specific issue and to a specific audience. In this section, I examine three trust relationships – epistemic trust, operational confidence and courtroom credibility – that bring together core issues, stakeholders and audiences. In terms of the audiences that stakeholders address, two types emerge as constitutive in these relationships. The first audience I focus on is derived from the notion of 'attentive publics' (Pierce and Lovrich, 1980), who hold an interest in the reliable development, use and evaluation of forensic technologies. The second audience is derived from the concept of 'publics-in-particular' (Michael, 2009) as those who are imagined to have a more immanent, active stake in the trust relationship, including other users of forensic information.

Epistemic trust

Trust in the underlying science of forensic genetics is often portrayed as a hard-won achievement. The forensic genetics community frequently explores accountability measures in the reporting of DNA analyses, such as the discussion of probability calculations:

> The error rates reported in this paper are useful for quality improvement and benchmarking, and contribute to an open research culture that promotes public trust. . . . Forensic DNA casework is currently regarded as one of the most important types of forensic evidence and has known many forensic success stories over the years. The basis of this is a combination of sound scientific principles, and a reliable and robust technological platform. . . . An error in forensic DNA analysis can lead to wrong decisions by investigative or legal authorities with far reaching consequences, such as conviction of innocent suspects, exoneration of guilty suspects, or failure to identify offenders. Furthermore, cases where a major miscarriage of justice was caused by an

> erroneous DNA result often generate a lot of media attention and damage the reputation of forensic laboratories.
>
> *(Kloosterman et al., 2014)*

Trust-building work undertaken by forensic geneticists consists of three recurrent tasks: to normalise interpretive intervals and error rates as part of statistical work, to acknowledge scientific limitations as a form of confidence in the science and a means to address these and to offer an avenue of addressing users of such technology as audiences. An essential part of all three is the comparison with other domains of data generation and application and aspired meaningful transparency about limitations such as error rates and their specific contexts – often portrayed as inherent to the scientific endeavour – that are safely manageable. This work also sets the scene for the precarious role of science in the operational context – the investigation – and, if successful there, also in the courtroom. Understanding matters of forensic fact in investigations and trials has become an integral part of forensic training literature directed at stakeholders who, while not researchers themselves, need to be able to understand the scientific basis of forensic reporting (Aitken et al., 2010; Good, 2001; Taylor et al., 2014; Forensic Science Service, 2005).

UK policy discourse echoes this linking of scientific culture and trust. In a speech to the Forensic Science Society on 8 November 2013, Andrew Millar, then chair of the UK Parliament's House of Commons Science and Technology Select Committee, reminded his audience:

> We need to have trust in forensic science and the way it is carried out. This is why regulation is so important. Private companies conduct forensic testing and, to some extent, crime scene activities, under a decent quality standards framework. But when a police laboratory is not working to those same standards, should the public trust the science they produce?

While Miller talked about public rather than stakeholder trust, I focus on his point about the need for a common quality standards framework. Miller indicated that trust in the delivery of forensic science services can be attained through the implementation of formal guidelines. Qualities such as comparability and reproducibility form the basis of trustworthy science in his speech. He discussed peer review and laboratory standards for the field. The audience he articulated is a recipient one that is less likely familiar with the scientific culture of forensic technologies but should be able to trust forensic findings when they are produced under a shared standard framework and as part of the CJS.

The setting up of the non-statutory Forensic Science Regulator (FSR) in the UK in 2007 was an effort to shore up epistemic trust in forensic science by advising the Home Office and forensic service providers on the validity and reliability of the underlying science of forensic technologies. Former FSR Andrew Rennison wanted 'the public and stakeholders to have confidence in the forensic science services provided to the CJS. To achieve this, forensic science must consistently meet high quality standards' (Forensic Science Regulator, 2013: 3). Rennison was concerned with two types of audiences: the broad recipient population of public goods arising from the use of forensic science and the particular stakeholders of the CJS, including the commissioners and users of forensic science such as the police and prosecution. Similarly, Rennison's successor, Gillian Tully, remarked on the value of standardised frameworks such as accreditation and best practice guidance for the epistemic trust relationship, with throughput impact on operational confidence and courtroom credibility. While reporting on the state of the art among forensic scientists, she was addressing a number of key stakeholders, in both policing and the CJS more broadly, as beneficiaries of developments in the epistemic trust relationship:

> More practitioners and experts now have objective evidence of their competence rather than a reliance on years in post or persuasiveness, neither of which is necessarily a good gauge of expertise. There is ongoing work to standardise interpretation of findings according to robust scientific principles, within the legal context of this jurisdiction. There are efforts to improve provision of proficiency tests, which provide comparative evidence of performance against peers and there is more collaboration between police forces, to help those who are lagging behind to catch up with implementation of quality standards.
>
> *(UK Forensic Science Regulator, 2021: 3)*

Such assertions are made in the context of concerns around science literacy, and thus, the reliability of expertise and capacity to work with data from forensic analyses (as well as expertise). Scientists who develop forensic applications, and who may also deliver casework, frequently articulate concern about the capacity of non-scientists to utilise and interpret information from forensic analysis, especially when it comes to understanding probabilities (Amorim, 2012: 265–266).

> People assume because it's DNA, other DNA based tests are going to come up with similar kinds of statistical certainty. And, of course, there's no guarantee that using DNA to look at one particular aspect of a person is the same as using DNA to prove identity or to match DNA samples between different locations.
>
> *(UK forensic geneticist, Respondent 1 [R1])*

Forensic scientists are often concerned about non-specialists' general lack of insight into the underlying science of genetic technologies (Tutton et al., 2014). This does reflect wider discourse in the public understanding of science, in particular the trope of educating non-scientists in order to help them realise the benefits of any given technology or scientific knowledge. While this may suggest a concern about lack of faith in science, forensic geneticists equally link lack of scientific insight to *over-confidence* in the capacity of technologies to deliver on criminal justice objectives.

> I think we have to deal with a number of misperceptions when we speak about public confidence. . . . [T]hey think that if a DNA profile is obtained

> then that means that somebody's guilty. . . . [N]ot only do the public think that but also some judges think that, a lot of lawyers think that, and certainly ministers think that as well.
>
> *(UK forensic geneticist, R2)*

This scientist's concern lies in the perceived inability of audiences of forensic genetics, both general and professional, to disentangle technological capacity and forensic analysis from interpretations and juridical fact-finding. Here, trust in the underlying science is separated from trust in the outcomes of criminal investigations and trials. Some of the reasons for such a narrative separation are associated with limited familiarity with the way forensic findings are produced. In this relationship of epistemic trust, attentive audiences are associated with interest in the contributions of forensic science to the production of common public goods such as security and justice. Publics-in-particular, including police, lawyers and other stakeholders, however, are actively engaged in the negotiation of forensically derived information where stakeholders' epistemic trust is ascribed material impact on the work of the CJS. I will come back to this point in the trust relationship of courtroom credibility.

Concern about trust in basic science resonates with other instances of imagined audiences of science (Besley and Nisbet, 2013; Irwin and Wynne, 1996; Michael and Brown, 2005; Petersen et al., 2009). Here, 'attentive publics' are often understood to be seen as relatively passive, whereas stakeholders as 'publics-in-particular' are considered to be key resources for producing legitimacy and support of forensic genetics in the CJS. Correlating the scientific basis of forensic genetics with claims about the field's capacity to deliver authoritative and contextually relevant forensic facts is a central concern in the negotiation of scientific standards as a source of trust. This is part and parcel of the notion of a forensic scientific culture (Cole, 2010, 2013; National Research Council, 1996) as a guarantor for reliable and useful forensic analysis. In the criminal justice context, forensic science stakeholders frequently seek to balance an acknowledgement of scientific limitation with the assurance that such limitations, for example statistical error rates, have no overall impact on the value of the science to the investigation. This narrative also holds that limitations can be managed via rational, scientific means such as standards and quality frameworks. This brings us full circle to Kloosterman and colleagues' 'open research culture' as guarantor of trustworthy forensic findings. In this narrative, achieving trust relies on the scientific underpinnings of forensic genetics alongside the rational management of forensic service provision. They work together as 'technologies of trust' (Leslie, 2010; Lewis and Atkinson, 2011; Porter, 1995).

Operational confidence

The second core trust relationship in forensic genetics focuses not on the underlying global scientific principles and standardised scientific methods, but on the shaping of this practice by local forensic and administrative technologies: in particular, the collection, preservation and analysis of biological materials recovered from crime scenes and suspects, as well as DNA profile searches of police databases. This relationship is informed by perceptions of policing uses of forensic technologies and the independence and integrity of forensic service provision in police work (Cole, 2013). Here, trust matters centre on the accountability of police and prosecution to other stakeholders and to society, within the context of established rationales for technology uses.

> There's a kind of bigger point to make about the political context and trust which is that one reason people are concerned about [function] creep is the creation of the national identity database which is shifting the idea of being able to check your identity using fingerprints from a situation where that's restricted to you being a suspect for a crime to a situation where it could happen potentially in a mandatory way to everybody and that's a big shift in values, whatever your position is on it, and that makes people less trusting of how decisions are being made about biometric information in general.
>
> *(UK forensic scientist, R3)*

This quotation links trust to the recognition that technologies are involved in the construction of social identities and relationships. It also suggests that such identities, as attributed to various audiences of uses of forensic technologies, may conflict with identities which those subject to technology use may construct of themselves. The operational remit of technologies becomes an issue of trust.

> There is a mixture of trust and distrust with the police. . . . [T]he people who have been arrested and not charged are a group who probably quite strongly distrust the police or tend to distrust the police more. There's going to be more distrust amongst those people that the [national DNA] database is going to be expanded, too . . . particularly if you think of the potential issues around race that you may be increasing distrust in a group of the population that it's widely recognised the police need to be increasing and building up trust with.
>
> *(UK senior police officer, R4)*

The ways in which law enforcement in the UK and elsewhere utilises forensic technologies are shown here to be central to some audiences' perceptions of forensic technologies use. The police officer suggests that mistrust in the fair and just application of technology can lead to mistrust in the technology itself. This is a significant trope in the negotiation of trust in forensic service provision and use across a variety of audiences. The forensic DNA database is a prime example in this narrative. In a conversation with the head of a police forensic laboratory in Germany, I discussed the technique familial searching (FS) as a fragile object of trust. FS describes the search of all profiles on a crime DNA database for a partial match with a crime scene trace in order to identify potential persons related to the unknown suspect for further investigation. The officer linked a lack of confidence in FS to diverse audiences' misinterpretation of its aims as well as the impact on persons becoming part of an investigation via the use of this technique. His main concern was that the DNA database as one type of forensic technology may suffer from a weakened trust relationship with some audiences because of the use of another technology, in this case FS. In order to avoid this, the officer suggested, such potential uses of the DNA database should be better communicated – he cited successful uses of FS in the Netherlands in order to strengthen operational confidence – and carefully regulated by law. Communication and legislation emerge as two key sources of operational confidence.

The establishment of the advisory National DNA Database Ethics Group (now Biometrics and Forensics Ethics Group) in 2007, as well as the introduction of the statutory office of the Biometrics Commissioner in 2013, are instances of building operational confidence in DNA technologies. Together with the Forensic Science Regulator, these institutions subscribe to integrity, impartiality and independence (and, ideally, effectiveness) as the guiding principles for forensic science provision in the UK. They are representative of the vital role of accountability in trust matters:

> I think this issue of accountability is key. . . . [P]eople on the database who were concerned have lost trust in the police because they've been arrested as a result of a false accusation.
>
> *(UK forensic scientist, R3)*

This linking of trust to accountability and public engagement imagines audiences' (mis)trust as understandable rather than elusive, as procedural rather than total – and, consequently, as producible. A variety of audiences are imagined, and the more specific such audiences become, the more pronounced imagined trust relationships are. The operational confidence narrative co-produces two types of audiences. Those who have been *subject* to technology use constitute the primary 'publics-in-particular'. The number and type of people who become part of this audience are growing with the expansion of what technologies can do and what they are used for, including forensic investigations and surveillance. Various narratives – on universal databases, drag nets, familial searching, 'big data' and genetic genealogy databases – intersect with the operational confidence relationship in the constitution of such specific audiences as stakeholders.

The most concise imagined audience of the operational confidence narrative is one that has seemingly lost confidence in the police use of forensic technologies because they were part of an investigation as suspects who turned out to be innocent. The moment of suspicion against a person is turned narratively into that person's suspicion of technology uses. Alternatively, audiences with positive experience are frequently articulated as those who will support forensic technology use based on their experience of successful outcomes (e.g., the solving of a cold case using DNA, the exoneration of a wrongfully convicted person). There is no persuasive data to support either assumption. The second audience is constitutive of the CJS itself, being comprised of those potentially subject to criminal justice processes. This audience is drawn upon to legitimise criminal justice uses of even not yet fully formed forensic technologies such as genetic age estimation or other phenotypical predictions. Some commentators argue that operational confidence must be negotiated in personal engagement with various stakeholders, including minorities, those processed by the CJS and specific neighbourhood populations. Overall, operational confidence relationships are portrayed as a policy and practice achievement of concern to any and all audiences, based on careful legislation and regulation and responsive to the strengths and limitations of technologies.

Courtroom credibility

This section focuses on the adversarial logic of the CJS in the UK and the USA. Some of the findings, however, can easily be applied across adversarial, inquisitorial and mixed forms.

A 2012 statement on forensic science by the US National Institute of Standards and Technology asserts that 'Public trust in the justice system relies on the validity and certainty of evidence presented to the courts' in order to help with 'reducing the number of mistrials or retrials related to questions about forensic analysis' (National Institute of Standards and Technology, 2012). This claim might be intuitively persuasive. Yet it does not engage with the fact that 'validity and certainty' are the outcome of court deliberations rather than properties of evidence that judges and advocates take on trust. In contested criminal trials that include forensic science expert testimony, prior assumptions of epistemic trust and operational confidence are suspended in favour of systematic efforts to test the credibility of experts and the opinions they offer to assist the work of the judiciary. Here, prior assumptions of scientific knowledge and standards as well as questions of evidence production and weight are all systematically deconstructed in the complex choreography of institutionalised mistrust (Wynne, 1989) that comprises the criminal trial in the adversarial system. Testimony given by expert scientists, evidence gatherers and others involved in chains of custody may all be contested and challenged and particular expert opinions included in, or dismissed from, the decision-making process. Criminal trials require assumptions of epistemic trust and assertions of operational confidence to be subject to critical examination in order to determine – beyond reasonable doubt – whether particular instances of forensic evidence, as well as their procurement and production, are accountable, legitimate and material to a determination of the facts in a case. The 2021 report by Gillian Tully, the recently retired FSR, reflects on these aspects as indicated in the earlier quotation and expands this as equally reliant on 'compliance with the Criminal Procedure Rules (CrimPR), Criminal Practice Directions (CrimPD) and other legal obligations' (Forensic Science Regulator, 2021: 23–24), suggesting that epistemic and operational trust relationships need to be enhanced by procedural awareness and compliance in order to build courtroom credibility: this trust relationship builds on the other two relationships as reflected in notions of chain of custody, leading to an image of courtroom credibility as greater than the sum of its constitutive parts.

Therefore, the achievement of courtroom credibility is a matter of negotiation and persuasion, based on testing the significance of evidential claims and the place of these claims in alternative narratives of the case. Within this process, attentiveness to wider audiences is suspended in favour of a focus on specific stakeholders within the criminal court as a 'public-in-particular'. This situated audience differs from other audiences that are elsewhere constructed as key collective actors in the trust landscape. Courtroom stakeholders in the UK are both participants in and witnesses to the dramaturgical examination and cross-examination of expert witness evidence, processes which contribute to an imperfect 'technology of trust' or, to be more precise, a technology of situated trustworthiness. Entangled within this 'technology' is expert witnesses' expertise, which is considered according to its capacity to provide epistemic and operational capacity to conduct reliable and credible forensic DNA analyses, as well as compliance with procedural legal regulations. It is at this point that the systemic and the personal dimensions of trust relationships overlap in the most apparent way. The technology is imperfect inasmuch as it is constitutive of a problematic process both pre-trial – if the judge is asked to consider admissibility (usually only relating to epistemic issues) – and during trial, when both epistemic and operational issues will be addressed and evaluated in a way that subordinates epistemic considerations to legal ones.

A 2012 UK Association of Chief Police Officers (now the National Police Chiefs Council) statement alludes to such contests over situated trustworthiness when it states that 'Any confidence gaps in science will be exploited to the full by the defence, therefore there is a need to ensure that these gaps are filled with reliable scientific methods' (Association of Chief Police Officers, 2012: 7). However, a paradox is created by the relevancies and practices of those involved directly in such contests which is not fully represented in this observation. It was laid bare in the Lord Chief Justice of England and Wales's 2014 Kalisher lecture, when he said:

> With increasingly complex or novel science there comes the risk of testing the science, rather than the evidence, in front of the jury. This in turns risks undermining juries' and public confidence in forensic science, with highly undesirable consequences, resulting either in less use of forensic evidence, or less use of juries. So there is a challenge for all of us – advocates and judges – to manage the presentation and testing of forensic evidence in such a way as to avoid fatally undermining confidence.
>
> *(Lord Chief Justice of England and Wales, 2014)*

These remarks reflect the long history in jurisprudence of discussing the nature and role of expert witnessing. Forensic scientists in general, and forensic geneticists in particular, are instances of this discussion. It is worth noting that there are formal ways in which the epistemic reliability and value of various kinds of genetic sciences and technologies are established to the satisfaction of courts (e.g., in the USA via Frye and Daubert standards, in the UK by dealing with admissibility as a preliminary issue). Yet confidence in the operational deployment of these technologies in the course of particular investigations and prosecutions is subject to different kinds of interrogation by a range of legal actors. Negotiation is an essential part of producing trust as credibility in the courtroom, as this intervention from a court-attending scientist further shows:

> There's a kind of a twitchiness or a nervousness coming from some of the prosecutors. . . . [T]hey were asking for a statistician to give the evidence. . . . [We] fear[ed] that more cases would fall because I wasn't a statistician, my colleagues weren't statisticians but we resisted that and convinced the prosecutors that in fact we could handle the cases, we could handle the evidence.
>
> *(Irish forensic scientist's account at a meeting, R8)*

The scientific witness here describes how expert credibility was re-asserted as an indispensable factor of trust in the presented evidence. The core challenge focused on the capacity of the witness to interpret scientific findings integral to the value of scientific evidence presented in hearings before trials and in court. This process illustrates the informal and semi-formal processes of testing trustworthiness as part of the formal trial process. It also reminds us that scientific credibility in the courtroom is related to the specific case as much as to the general qualities attributed to the forensic science resource in question.

Discussion

Borrowing Georg Simmel's understanding of trust as being a 'highly variable, relational process' (Frederiksen, 2012: 734) and Niklas Luhmann's (1979) concern with the stability of systems, in this contribution, I have interrogated trust issues as they arise and are reconfigured in the course of the development and application of forensic genetics in the laboratory, the criminal investigation and the trial. I have explored the variety of audiences addressed in these differing contexts and how they are woven into assertions about forensic genetics as a technology of situated trustworthiness. The ways in which trust concerns in social interactions play out within the broader systemic preoccupation with trust in forensic genetics provides an example of the translation from trust in personal relationships to trust in a system of institutionalised relationships (Derksen, 2010).

Different forms of technolegal worlds are reflected on in each of the three trust relationships – epistemic trust, operational confidence and courtroom credibility. Simultaneously, these worlds ideally align in the fact-finding process in trial. I emphasise the ideal nature of such a technolegal world-coming-together because the trust relationships in science and in policing (as the operational use of science) do not necessarily lead to courtroom credibility; rather, they depend on the dramaturgical rules of the courtroom. The negotiation of trust, together with the invocation of people who are narrated to have a key role and stake in the information from and negation of forensic genetics data, materialise trust relationships that are of investigative (policing) and legal (courtroom) relevance. These relationships are the loci at which forensic data, scientific claims, the operational use of forensic technologies and their value as investigative intelligence or as trial evidence gain ontological substance in and for an investigation, the prosecution and judiciary fact-finding processes (see also Toom, 2016).

The specific types of audiences to trust invocations who are part of trust relationship can be said to represent a network of accountability and legitimacy. Broader audiences are posited as principals by those responsible for the reliable, safe and, arguably, efficient delivery of forensic genetics. Others operate as agents to those principals, who interpret and deliver specific tasks and responsibilities delegated to them in the CJS (Gallo, 2009; Guston, 1999). Here, the principals are discursively constituted by the agents in enacting a state of trust or confidence, usually in the form that a trust giver can rely on the agent to make appropriate judgments about the trust relationship. Each of the three discussed trust relationships is taken as representative of an aspect of the CJS, but each focus on specific matters of concern: scientific foundation, operational use and dramatic evaluation. These narratives are concerned with achievements communicated to particular publics: epistemic trust to technology publics and criminal justice stakeholders, operational confidence to those actually and potentially processed by the CJS, and juridical credibility to the attentive public of the trial (see Table 9.1). Systemic trust is enacted in the agent's definition of trust to which the co-produced particular publics are encouraged to subscribe. The trajectory of these trust narratives is orientated towards legitimising the outputs of the CJS, rather than its inputs. This feature is shared across different logics of the CJS (i.e., adversarial, inquisitorial or mixed forms). Whereas in the adversarial(-leaning) system trust is negotiated with the judge in determining what evidence is admissible as well as in courtroom proceedings, in the inquisitorial(-leaning) system, this trust is foremost negotiated in evidence reports and their evaluation by the judiciary, and here, too, scientists can be called as expert witnesses to

TABLE 9.1 Trust relationships in forensic genetics

Relationship	*Emphasis*	*Audiences*
Epistemic trust	The underlying science of forensic technologies	'Attentive publics': those expecting forensic science to contribute to security and justice 'Publics-in-particular': those working with information from forensic analysis
Operational confidence	The use(s) of forensic applications, especially by agencies of the state	'Attentive publics': those potentially benefiting or being subject to use 'Publics-in-particular': those with experience of being subject to forensic analysis
Courtroom credibility	The testing of scientific and operational aspects of forensic science in juridical fact-finding	'Attentive publics': generally not articulated 'Publics-in-particular': key stakeholders in courtroom proceedings (e.g., judge, jury, defence)

continue the negation of trust – and materialise relevance in a trust relationship – during a trial.

References to trust, confidence and credibility and the construction of a variety of audiences embody performative elements: assertions of the presence or absence of forms of trust are usually made to achieve specific ends rather than to provide disinterested descriptions. Using the language of trust invokes expectations about relationships, attempting to enroll others in certain behaviours. The speech act of invoking trust can operate as a way of shoring up expertise, of assigning or even claiming responsibilities, of promoting support or of urging change and of articulating wider (recipient) audiences and (involved) stakeholder groups (Szerszynski, 1999). Rather like Simon Cole's (2010) analysis of 'science as work' as 'dividing forensic labour in a set of general tasks', the enactment of public trust and their 'attentive' and 'particular publics' – the identity work involved in performing what trust is and in which relationships it exists – is an active process of work. Audiences of trust are a strategic device in the negotiation of the legitimacy and, essentially, the value of forensic science to criminal justice. Equally, assertions of lack of trust provide opportunities to challenge the uses of forensic genetics and also to lever changes in the ways that forensic genetic information is made and consumed within the CJS. Because they are fragile, these trust relationships need to be produced and reproduced continually. The co-articulation of trust (plus its semantic variations) and audiences of trust is active work in maintaining legitimacy for the use of science in law. Trust is applied as a parameter in discursively maintaining the stability of the CJS.

Concluding remarks

Trust relationships rely on trust as 'social glue' and as 'procedural oil'. Trust is therefore a placeholder for a variety of complex knowledges, processes and regulations on which stakeholders in the CJS need to be able to rely – but not without contestation. The establishment and maintenance of trust in forensic genetics involves ongoing negotiation between key actors –scientists, investigators, judges – and their invoked audiences, including specific stakeholders such as users of forensic data in epistemic trust relationships, suspects and others in operational confidence and juries and society in courtroom credibility, over the meaning of forensic genetic knowledge, expertise, technologies and methods relevant to the working of the CJS and its processes. Trust is continually constructed and contested between those calling upon it and the devices, methods and technologies of interpretation they use. There is a complex contexture at work here in which views of a variety of stakeholders, knowledges, organisations, desires and fears, as well as a variety of political imperatives, are in play. Key trajectories in which trust issues appear and are dealt with in the CJS significantly differ from those in other fields of genetic science because of the ways in which claims to scientific expertise can be contested within the court setting but also in the operational setting where investigators decide if forensic information is useful or the outcome of an investigation shows whether the use of forensic data by investigators can be trusted to lead to detection. Furthermore, the detailed ways in which references to trust frequently characterise identities and relationships remind us that our focus should not only be on trust in science and policing but also on the ways in which scientific, policing and judicial actors enact trust relationships with each other and with the stakeholders and wider audiences whom they serve.

Trust is a systemic property of the CJS rather than a qualifying one: trust is invoked and produced, but it is not measured, nor does it necessarily persist across trust relationships of the CJS. The three discussed trust relationships reflect on technolegal worlds in which forensic genetics is given the quality of a technology of situated trustworthiness.

References

Aitken, C., Roberts, P. and G. Jackson (2010) *Fundamentals of Probability and Statistical Evidence in Criminal Proceedings. Guidance for Judges, Lawyers, Forensic Scientists and Expert Witnesses.* London: Royal Statistical Society.

Amorim, A. (2012) Opening the DNA Black Box: Demythologizing Forensic Genetics. *New Genetics and Society* 31(3): 259–270.

Association of Chief Police Officers (2012) *Live-Time Forensics. Harnessing Science and Innovation for Forensic Investigation in Policing.* Harrogate: ACPO.

Barbalet, J. (2006) *A Characterization of Trust and its Consequences.* Leicester: Social Contexts and Responses to Risk Network.

Besley, J.C. and M. Nisbet (2013) How Scientists View the Public, the Media and the Political Process. *Public Understanding of Science* 22(6):644–659.

Cole, S.A. (2010) Acculturating Forensic Science: What is 'Scientific Culture', and How Can Forensic Science Adopt It? *Fordham Urban Law Journal* 38(2):435–472.

Cole, S.A. (2013) Forensic Culture as Epistemic Culture: The Sociology of Forensic Science. *Studies in History and Philosophy of Science Part C: Studies in History and Philosophy of Biological and Biomedical Sciences* 44(1):36–46.

Derksen, L. (2010) Micro/Macro Translations: The Production of New Social Structures in the Case of DNA Profiling. *Sociological Inquiry* 80(2):214–240.

Engdahl, E. and R. Lidskog (2014) Risk, Communication and Trust: Towards an Emotional Understanding of Trust. *Public Understanding of Science* 23(6):703–717.

Forensic Science Regulator (2013) *Business Plan 2013/14.* London: Forensic Science Regulator.

Forensic Science Regulator (2021) *Annual Report 2019–2020.* London: Forensic Science Regulator.

Forensic Science Service (2005) *Guide to DNA for Lawyers and Investigating Officers.* Birmingham: Forensic Science Service.

Frederiksen, M. (2012) Dimensions of Trust: An Empirical Revisit to Simmel's Formal Sociology of Intersubjective Trust. *Current Sociology* 60(6):733–750.

Gallo, J. (2009) The Discursive and Operational Foundations of the National Nanotechnology Initiative in the History of the National Science Foundation. *Perspectives on Science* 17(2):174–211.

GeneWatch UK (2010) *National DNA Database. Submission to the Home Affairs Committee.* Buxton: GeneWatch UK.

GeneWatch UK (2011) *GeneWatch UK Briefing: DNA Databases and Human Rights.* Buxton: GeneWatch UK.

Good, P. (2001) *Applying Statistics in the Courtroom: A New Approach for Attorneys and Expert Witnesses*. Boca Raton: Chapman & Hall/CRC.

Guston, D.H. (1999) Stabilizing the Boundary between US Politics and Science: The Role of the Office of Technology Transfer as a Boundary Organization. *Social Studies of Science* 29(1):87–111.

Haerlin, B. and D. Parr (1999) How to Restore Public Trust in Science. *Nature* 400:499.

Home Office (2013) Press Release: Biometrics Commissioner Appointed. (4 March), *Home Office*. Available at: www.homeoffice.gov.uk/media-centre/press-releases/biometrics-commissioner-appntd (accessed 6 September 2017).

Hough, M. (2021) Trust and Legitimacy. The Basic Ideas. In: *Good Policing. Trust, Legitimacy and Authority*. Bristol: Policy Press, pp. 13–30.

House of Lords Science and Technology Select Committee (2019) *Forensic Science and the Criminal Justice System: A Blueprint for Change (3rd Report of Session 2017–19)*. London: The Stationary Office.

Human Genetics Commission (2009) *Nothing to Hide, Nothing to Fear? Balancing Individual Rights and the Public Interest in the Governance and Use of the National DNA Database*. London: Human Genetics Commission.

Irwin, A. and B. Wynne (eds.) (1996) *Misunderstanding Science? The Public Reconstruction of Science and Technology*. Cambridge: Cambridge University Press.

Jacob, M. and T. Hellström (2000) Policy Understanding of Science, Public Trust and the BSE-CJD Crisis. *Journal of Hazardous Materials* 78(1–3):303–317.

Jasanoff, S. (1995) *Science at the Bar: Law, Science and Technology in America*. Cambridge, MA: Harvard University Press.

Kaye, D.H. (2010) *The Double Helix and the Law of Evidence*. Cambridge, MA: Harvard University Press.

Khodyakov, D. (2007) Trust as a Process: A Three-Dimensional Approach. *Sociology* 41(1): 115–132.

Kloosterman, A., Sjerps, M. and A. Quak (2014) Error Rates in Forensic DNA Analysis: Definition, Numbers, Impact and Communication. *Forensic Science International: Genetics* 12:77–85.

Leslie, M. (2010) Quality Assured Science: Managerialism in Forensic Biology. *Science, Technology & Human Values* 35(3):283–306.

Lewis, J. and P. Atkinson (2011) The Surveillance of Cellular Scientists' Practice. *BioSocieties* 6(4):381–400.

Lidskog, R. (1996) In Science We Trust? On the Relation Between Scientific Knowledge, Risk Consciousness and Public Trust. *Acta Sociologica* 39(1):31–56.

Lord Chief Justice of England and Wales (2014) Expert Evidence: The Future of Forensic Science in Criminal Trials. The 2014 Criminal Bar Association Kalisher Lecture. (14 October), *Judiciary*. Available at: www.judiciary.uk/wp-content/uploads/2014/10/kalisher-lecture-expert-evidence-oct-14.pdf (accessed 16 May 2019).

Luhmann, N. (1979) *Trust and Power*. New York: Wiley.

Lynch, M., Cole, S.A., McNally, R. and K. Jordan (2008) *Truth Machine. The Contentious History of DNA Fingerprinting*. Chicago: The University of Chicago Press.

M'charek, A., Hagendijk, R. and W. De Vries (2013) Equal Before the Law: On the Machinery of Sameness in Forensic DNA Practice. *Science, Technology & Human Values* 38(4):542–565.

Metropolitan Police Authority Civil Liberties Panel (2011) *Protecting the Innocent. The London Experience of DNA and the National DNA Database*. London: Metropolitan Police Authority.

Michael, M. (2009) Publics Performing Publics: Of PiGs, PiPs and Politics. *Public Understanding of Science* 18(5):617–631.

Michael, M. and N. Brown (2005) Scientific Citizenships: Self-representations of Xenotransplantation's Publics. *Science as Culture* 14(1): 39–57.
Möllering, G. (2001) The Nature of Trust: From Georg Simmel to a Theory of Expectation, Interpretation and Suspension. *Sociology* 35(2):403–420.
Molyneux-Hodgson, S. and M. Meyer (2009) Tales of Emergence—Synthetic Biology as a Scientific Community in the Making. *BioSocieties* 4(2–3):129–145.
National Institute of Standards and Technology (2012) Congressional and Legislative Affairs. FY13: Measurement Science and Standards in Support of Forensic Science. (9 February), *National Institute of Standards and Technology*. Available at: www.nist.gov/public_affairs/factsheet/forensic-science2013.cfm (accessed 6 September 2017).
National Research Council (1996) *The Evaluation of Forensic DNA Evidence*. Washington, DC: National Academy Press.
O'Neill, O. (2004) Accountability, Trust and Informed Consent in Medical Practice and Research. *Clinical Medicine* 4(3):269–276.
Petersen, A., Anderson, A., Allan, S. and C. Wilkinson (2009) Opening the Black Box: Scientists' Views on the Role of the News Media in the Nanotechnology Debate. *Public Understanding of Science* 18(5):512–530.
Pierce, J.C. and N.P. Lovrich Jr. (1980) Belief Systems Concerning the Environment: The General Public, Attentive Publics, and State Legislators. *Political Behavior* 2(3):259–286.
Porter, T.M. (1995) *Trust in Numbers. The Pursuit of Objectivity in Science and Public Life*. Princeton, NJ: Princeton University Press.
Prainsack, B. (2010) Key Issues in DNA Profiling and Databasing. Implications for Governance. In: Hindmarsh, R. and B. Prainsack (eds.) *Genetic Suspects. Global Governance of Forensic DNA Profiling and Databasing*. Cambridge: Cambridge University Press, pp. 15–39.
Roberts, P. (2013) Renegotiating Forensic Cultures: Between Law, Science and Criminal Justice. *Studies in History and Philosophy of Science Part C: Studies in History and Philosophy of Biological and Biomedical Sciences* 44(1):47–59.
Skinner, D. and M. Wienroth (2019) Was This an Ending? The Destruction of Samples and Deletion of Records from the UK Police National DNA Database. *BJHS Themes* 4:99–121.
Szerszynski, B. (1999) Risk and Trust: The Performative Dimension. *Environmental Values* 8(2):239–252.
Taylor, D., Bright, J.A., Buckleton, J. and J. Curran (2014) An Illustration of the Effect of Various Sources of Uncertainty on DNA Likelihood Ratio Calculations. *Forensic Science International: Genetics* 11:56–63.
Toom, V. (2016) Whose Body Is It? Technolegal Materialization of Victims' Bodies and Remains after the World Trade Center Terrorist Attacks. *Science, Technology, & Human Values* 41(4):686–708.
Tutton, R., Hauskeller, C. and S. Sturdy (2014) Suspect Technologies: Forensic Testing of Asylum Seekers at the UK Border. *Ethnic and Racial Studies* 37(5):738–752.
Williams, R. and P. Johnson (2008) *Genetic Policing: The Use of DNA in Criminal Investigations*. Cullompton: Willan Publishing.
Wynne, B. (1989) Establishing the Rules of Laws: Constructing Expert Authority. In: Smith, R. and B. Wynne (eds.) *Expert Evidence: Interpreting Science in the Law*. London: Routledge, pp. 23–55.
Wynne, B. (2006) Public Engagement as a Means of Restoring Public Trust in Science—Hitting the Notes, But Missing the Music? *Community Genetics* 9:211–220.

10

WHY IS DNA NOT ENOUGH?

The multiple temporalities of family reunification in Finland

Anna-Maria Tapaninen and Ilpo Helén

Introduction

After the discovery of genetic fingerprinting in 1985, DNA analysis was first applied not in criminal investigations but to verify the identity of a migrant child in the UK via analysis of his genetic relatedness to his mother and siblings (Jeffreys et al., 1985). Thereafter, since the 1990s, DNA testing has become an established method for family reunification and is currently used in at least 25 countries in the global North, including 21 countries in the EU (EMN, 2016, 2017). As family reunification has been one of the main avenues of immigration to the EU for the past 20 years, amounting to 28% of all first residence permits issued to third-country nationals in 2018 (Eurostat, 2019), the numbers of DNA tests – which are seldom disclosed – are most likely considerable. In Finland, the proportion of family migration out of all residence permits was 35% in 2018 (EMN, 2019).

In administrative procedures, DNA analysis is used when the documents formally establishing family relations, such as birth or marriage certificates, are either missing or deemed unreliable. Even though this rationale is apparently simple, the procedures are enmeshed in complex and varying regimes of immigration management. In practice, DNA testing mostly concerns people with a refugee background – i.e., humanitarian migrants. This context makes a significant difference in investigations conducted by the immigration authorities. In Finland, the methods used in such cases include lengthy interviews and, often but not always, DNA testing.

The ambivalence concerning DNA proof was captured by one of our interlocuters, a Finnish Somali, who repeatedly posed a rhetorical question during our interview with him: 'Why is DNA not enough?' Another starting point for our study was a press release by the Finnish Immigration Service (henceforth, Migri) from 2008, in which the limited role of DNA testing was concisely clarified: 'A purely biological relationship is not, however, sufficient for a positive decision on a

DOI: 10.4324/9780429322358-14

resident permit without the background of a genuine, permanent family life'. We will return to this statement of the veracity of family ties throughout the chapter. Instead of deciphering the consequences of forensic DNA testing, we simply ask, 'Why is DNA not enough?' To answer this question, we will look at *how* the family is enacted in compiling – or, rather, making – evidence of family ties. Our analysis weaves together two angles: temporalities and kinship. This analytical fabric, we think, will make evident the ambivalences of the scene.

Time and relations in a technolegal regime of family reunification

Although there are a few comparative studies on the deployment of DNA analysis for family reunification (Taitz et al., 2002; Villiers, 2010; Holland, 2011; La Spina, 2012; Heinemann et al., 2013; Dove, 2013 Heinemann et al., 2015; Granados Moreno et al., 2017; Lee & Voigt, 2020), the actual weight of DNA evidence remains unclear. DNA testing, despite its perceived accuracy and objectivity, is inherently equivocal, which is particularly evident in the case of Finland (see Helén & Tapaninen, 2013; Tapaninen & Helén, 2015, 2020; Tapaninen et al., 2019; Halme-Tuomisaari et al., 2019). In addition to the differences between countries, the significance of DNA testing also varies with the labile regimes of immigration management, which, in turn, respond to the shifting trends in migration (e.g., Heinemann et al., 2015; Tapaninen & Helén, 2020). In short, the rationale for DNA testing for family reunification regularly changes – probably everywhere – despite the established technical procedures. Therefore, the temporalities of the migration apparatus are pertinent to our analysis.

The weight of time is critical in forced migration. A great majority of asylum seekers have fled from chaotic and precarious circumstances, and their kinship practices are in flux too. People flee, lose contact, die, disappear, are widowed or orphaned, join new households and try to join their kin living elsewhere. Old and new relations provide essential resources of information, money and shelter during and after the journeys. People's family life entails an inevitable 'thickening and thinning of relatedness' (Carsten, 2013) as an integral part of kinship practices. These plights and arrangements are recurrent in the decisions of the Administrative Court of Helsinki we analysed. These examples bring to the fore both the vicissitudes of time in forced migration and the parallel weight of kinship.

The scenery sketched here depicts some of the complications encountered by decision makers at Migri. This essentially outlines the landscape of our analysis. In contrast to the applicants with reliable documents, the 'recipients of international protection' seldom have such documents, and therefore, their relations and identities are harder to decipher. Immigration authorities try to tackle this complexity by various biometric methods. Among them are technologies that seek 'the truth from the body' (Fassin & Halluin, 2005), including fingerprinting (van der Ploeg, 1999), medical age assessment (Tapaninen, 2018; Netz, 2019), medical examination of signs of torture or affliction (Fassin & Halluin, 2005; Ticktin, 2011) and

DNA analysis. These methods display a repertory of entwined 'technologies of doubt and proof' (Good et al., 2015). It is evident that the quest for precise 'truth' also intensifies ubiquitous suspicion by immigration authorities (see Tapaninen & Helén, 2015; Hall & Naue, 2015). DNA analysis for family reunification is one of these widespread biotechnological tools yet is specific because it concerns relational identities and not solely the identification of individuals. Therefore, it is an utterly reductive way of imagining, defining and verifying relatedness.

In this chapter, we problematise the views that DNA analysis is a key factor leading to a geneticised notion of the family (e.g., Heinemann & Lemke, 2013, 2015). DNA analysis does not do anything by itself. If and when it does, the objects of analysis – family ties, kinship or credibility – are enacted (see Mol, 2002) via shifting relations between law, bureaucracy and technoscience (e.g., Halme-Tuomisaari et al., 2019; Tapaninen & Helén, 2020). Many STS studies have shown that biotechnological evidence is relational, context dependent and often contested in the context of immigration management (e.g., Tapaninen, 2018; Netz, 2019; Netz et al., 2019) and more generally (e.g., M'charek, 2000; Kruse, 2015; M'charek et al., 2020). Our analysis here continues this line of study: we analyse the interlacing of legislation, administration and biotechnology around family reunification in Finland as a technolegal regime. This regime is central in processes that make up 'families' in immigration control (see Tapaninen & Helén, 2020); within it, DNA analysis is used, and the results of the tests are interpreted and given evidential weight in varying ways by officials, experts and people subjected to scrutiny (see Introduction in this volume; Kruse, 2015).

Much of the existing literature on DNA testing for family reunification focuses, for obvious reasons, on the definition of 'family' and especially on the dichotomy between the 'biological' and the 'social' meaning of *families*. In this chapter, we adopt a different approach and problematise the dualism. We largely leave the predominant dichotomy aside as the notion of 'social family' is utterly abstract and formless. However, this dualism is constitutive for the rationales of family reunification and its technolegal apparatus. Therefore, the dualism is part of our empirical analysis.

We suggest that the concept of kinship provides a fruitful and even obvious perspective for the study of DNA analysis within the technolegal regime of family reunification (see Haimes & Toom, 2014).[1] Kinship permeates not only migrancy but also the migration regime. After all, the right to family, as laid down in myriad international conventions and national legislations, illustrates the persistence of kinship at the heart of modernity (see McKinnon & Cannell, 2013). DNA analysis is basically about generational succession – that is, about genetic kinship. In administrative procedures, the object of investigation is 'family', but knowledge of this entity is interwoven with examining and doubting 'messy' kinship practices. When defining eligible 'families', immigration officials categorise and regulate peculiar forms of relatedness, which, in their complexity, reflect the globalisation of kinship dynamics (e.g., Ginsburg & Rapp, 1995; Browner & Sargent, 2011). Not only are the relatives scattered around the world, but the global connections are also enacted by DNA testing as a significant part of the technolegal regime.

We amend our discussion of kinship with an analysis of the temporalities of family reunification processes. With the help of this combination, we look more closely at the ambivalences of the technolegal regime of family reunification. Time unfolds in multiple ways, both in migration trajectories and in the migration apparatus. The passing of time in the sense of duration is in many ways fundamental to the process of family reunification. First, kinship is about time in the sense of enduring ties, yet it is versatile practices and strategies for making and maintaining alliances that fold these ties (e.g., Bourdieu, 1977). Second, as most of the sponsors in such applications have a refugee background, their trajectory has been suffused with (im)mobility (Lems & Tošić, 2019) on the way to and throughout these bureaucratic proceedings. This predicament has been framed as prolonged waiting or 'stuckedness' (Hage, 2009; Cabot, 2012; Griffiths, 2014; Jacobsen et al., 2020) and has been explored in many evocative ethnographies combining descriptions of the subtle tactics of border crossers with analysis of border regimes (e.g., Khosravi, 2010; Feldman, 2011; Griffiths, 2014; M'charek, 2020; Razakou, 2020). We find Andersson's (2014) discussion on the coming together of the bureaucratic time of control and the contingent time of migration especially fruitful. To his concepts, we add what we call the time of legislation and the time of technoscience. The temporalities of family reunification have not been examined widely (for the Finnish context, though, see Leinonen & Pellander, 2020; Näre, 2020). In most analyses of DNA testing for family reunification, time has hardly featured, and thus the analysis has tended to be quite abstract (but see Tapaninen et al., 2019). Our contribution seeks to address this lacuna. We believe this analytical lens focused on temporalities within a technolegal regime can make visible the indeterminacy and variability of immigration control.

In this chapter, temporalities entail first the time of legislation. By this, we refer to the changing provisions that directly and indirectly define the role of DNA analysis for family reunification. In addition, we look at the temporal implications of the selected provisions. Second, the time of forensic DNA testing addresses technoscientific temporality, opening up a largely self-referential and self-vindicating world. Third, the bureaucratic time is closely linked to temporal orders of law and forensic genetics, yet administrative practices and procedures form a specific temporality where the criteria of eligibility are played out. We will not discuss the time of migration in a separate section but intertwine it with the other temporal frames.

Our analysis is based on legislative, administrative, policy and scientific documents and about 60 interviews with immigration officials, lawyers, geneticists, experts in NGOs and people subjected to the family reunification interrogation in Finland. In addition, we have analysed 253 decisions of the Administrative Court of Helsinki between 2003 and 2014 that include the term *DNA*. This data was collected in three research projects between 2010 and 2020[2] and thus also spans the period of changing legislation and related variations in administrative procedures.

The temporalities of and in legislation: The rule of law and elusive regulations around DNA testing

In Finland, the temporal nexus between immigration and technolegal regimes is particularly clear. The year 1990 is perceived as a turning point with the arrival of Somali asylum seekers in a country where a very small number of immigrants had arrived since the Second World War. Soon afterwards, these asylum seekers' applications for family reunification wrought new challenges, and to solve the problem of rejected and pending cases, the Directorate of Foreigners – the predecessor of Migri – together with ministries, geneticists and NGOs examined the potential of DNA testing in Eastern Africa in 1996–1997. Contemporary documents and our interviews indicate that the project was driven not only by helplessness and distrust but also by the need to secure the applicants' rights (Tapaninen & Helén, 2015). As a result, the Aliens Act was amended with two sections on DNA analysis that came into force in 2000, whereby Finland was the second country in Europe, after Denmark (see Olwig, 2020), to incorporate DNA testing into migration legislation.

In this section, we begin our discussion of the temporality of the Finnish family reunification procedures by taking the law as our point of departure. We focus on two dimensions: the temporal context of the legislative amendments and the figuration of time in the definitions of eligibility. As we will show, the technolegal regime reaches beyond the two sections of the Aliens Act on DNA testing and also encompasses other provisions of the act regulating family reunification. Even though the provisions on DNA testing have largely been unaltered for two decades, save for some technical details (see the following section), the new Aliens Act of 2004 and numerous amendments since have changed the scope of DNA testing too.

Since its introduction, consensus about DNA analysis for family reunification has been conspicuous in Finland, and it was guaranteed by essential legal principles. Testing was to be not only voluntary but also free of charge for the people to be tested. Pursuant to Section 65 (2) of the Aliens Act, people tested were also to give their 'written consent based on information and free will'. Furthermore, under Section 66 (1), the use of the data is restricted as 'the samples and the data concerning DNA identification shall be destroyed' after the decision is made. These two routines have legitimated the use of biometrics for family reunification (Tapaninen & Helén, 2015). A sort of official complacency was also echoed in the headline of the press release quoted in the introduction (Finnish Immigration Service, 2008): 'DNA analysis has unified a record number of immigrant families'.

From the beginning, there has been a vacillation between geneticised and degeneticised notions of relatedness and thus a constant tension between the social and biological ideas of kinship. The provision on 'establishing family ties by means of DNA analysis' in Section 65 (1) encapsulates the basic principles that also establish the inherent ambiguity of DNA testing:

> The Finnish Migration Service may provide an applicant or sponsor[3] with an opportunity to prove their biological kinship with DNA analysis paid from

> state funds if no other adequate evidence of family ties based on biological kinship is available and if it is possible to obtain material evidence of the family tie through DNA analysis.

Hence, DNA testing is an option and not a demand, let alone a compulsory requirement. The section states that the option may or may not be given by Migri. Further, the people concerned do not have to cover the costs, but neither can they organise or demand testing themselves, which partly explains the significant number of appeals to the court where DNA is mentioned (see later in this chapter). Importantly, Section 65 refers to biological kinship but simultaneously curtails the importance of DNA evidence. The last sentence implies that DNA analysis shall only be a last resort. It also provides support for the principle we pointed out in the introduction: DNA evidence is not enough if there is insufficient evidence of the veracity of the family tie. Then, it is up to the administration and appellant courts to decipher what counts as sufficiently permanent and convincingly genuine family ties in circumstances suffused with insecurity and separation, partly produced by the migration apparatus itself.

The provisions of DNA testing have been practically unaltered for two decades, yet other amendments regulating family reunification have had momentous consequences for the weight of DNA analysis. This is closely related to the restrictive turn in legislation and policy since the late 2000s.[4] This trend clearly demonstrates how, to quote Kelly (2015: 184), 'doubt is a social process forever on the move'. The Aliens Act was amended in 2010 with the explicit goal of 'cutting down certain pull factors' (HE 240/2009: 4). Six of the nine amendments directly or indirectly concern family reunification. In general, the purpose was to fight fraud via more reliable proof of identities and relations. Proof of the latter, in turn, made time in the sense of proven 'permanence' a significant criterion.

While the temporal indications are mostly implicit in the Aliens Act, time does count in indirect ways.[5] This has been particularly clear in the case of more 'irregular' relations – that is, ones that go beyond the bounds of stable nuclear families. As the definition of eligible kin is, in principle, quite extended in the Aliens Act, these cases have not been rare. A conspicuous case is foster children. In many cases, such people are nieces, nephews or grandchildren, and therefore, their relatedness to the sponsor can be verified by DNA analysis even though the legislation only refers to 'guardians'. Nevertheless, the care given before the departure of the sponsor for Finland and the need for such care at the time of the application are crucial temporal requirements. In other words, DNA analysis and indications of family life continuity co-exist as important evidence of veritable family ties; in decision making over family reunification, it is the latter evidence that folds the past and the presumed future into the present, not the former (cf. Harbers, 2005). We discuss this topic more in the section on bureaucratic time.

In short, the restrictive turn of immigration policy diminished the role of DNA analysis as there were new and more complex issues to consider (Tapaninen & Helén, 2015). In 2011, an immigration lawyer summarised the shift as follows:

'DNA was at the centre when it was first applied, it was the definitive factor, and it was extremely easy and comfortable for everybody because it was so clear'. She remarked that after the tightening of the criteria, a negative decision can be based 'on grounds that can be made up'. Later in this chapter, we analyse how immigration authorities examine case by case whether family life has been permanently stable enough or not and whether it is supposed to continue in the future, despite separation.

More legal constraints for family reunification were yet to come – more efficient ones. In response to the 'refugee crisis' of 2015, the Finnish government swiftly tightened the requirements for family reunification. The question was less about the biotechnological or narrative proof of veritable families than about money and time (see Näre, 2020). Most importantly, the government started to demand a considerable 'stable income'. According to the new rules, a family of two adults and two minors should have a monthly net income of at least €2,400, which exceeds the median income in Finland. Only refugees are exempt from this requirement – but only if an application is filed recognized within three months of the date the sponsor was granted his or her refugee status. These demands compel people to chase time and invest in resources such as income, information and the capability to navigate the bureaucratic maze, often with the help of kin. Simultaneously, meticulous interviews and DNA testing are still actively deployed as methods of family reunification management, as an immigration officer recently told us. In fact, a record number of DNA tests were conducted in 2019.[6]

The temporality of technoscience: Snapshots out of time

Research on DNA analysis for family reunification usually focuses on immigration control that may lead to the cutting of (social) family ties (e.g., Dove, 2013; Heinemann & Lemke, 2013, 2015) – i.e., on legal and administrative containment via science – and the role of laboratories is often highlighted. It is as if DNA testing were the decisive factor leading to the rejection or acceptance of applications. In Finland, the scientific validity of DNA analysis has not been contested. Since the beginning, the markers used have multiplied, and today, with 23 markers, there is little need for data on reference populations.[7]

In Finland, the people to be tested have minimal direct contact with the laboratories: they cannot instigate the testing or contact the laboratories themselves. Migri offers the option of DNA testing, co-operates with the embassies and the laboratory of the National Institute of Health and Welfare,[8] receives the results and makes a decision. Since the recent amendments to the Aliens Act in 2019, a buccal swab sample has been taken by the people concerned themselves or by an official. There is no need to go to the laboratory in person. The result is thus plainly built on 'a relationship of genetic relatedness between at least two *samples*' (Anderlik & Rothstein, 2002: 215, emphasis added).

From this perspective, the result is a snapshot-like representation and thus far removed from the kinship understood through practises unfolding (Bourdieu,

1977). DNA analysis is built on time in the sense of generational succession – i.e., on the passing of genes between generations. In the context of family reunification, it performs a specific time order, astonishingly different from 'the folding of past and future into the present' (Harbers, 2005: 359) occurring in forensic DNA analysis in the context of the criminal justice system. In family reunification cases, the report of DNA analysis expresses the result in numbers that indicate probability. This representation is out of time as well as disembodied and detached from the actual making and unmaking of relatedness. This apparent timelessness of DNA analysis[9] stands in stark contrast to the contingent time of migration and the unpredictable bureaucratic time.

A geneticist encapsulated the distinctiveness of the technoscientific world by saying that he and his colleagues 'do not want to know what had happened before the sample arrived in the lab or what happens afterwards'. From the perspective of geneticists, the truth of their analyses is established by probabilities that do not add up to a straightforward rejection or confirmation. Nor do the experts refer to families; rather, they refer to genetic relatedness or kinship. In Migri, instead, the results are translated into evidence of the existence of families. At this point, though, the proof does not lead to a definite conclusion, for the principle of veracity we underlined earlier does make a difference: immigration authorities demand additional evidence of 'genuine' and 'permanent' family life.

In the fact sheet provided by Migri (Finnish Immigration Service, 2019) in several languages, the basics of DNA analysis are explained as a background for informed consent:

> DNA analysis is used to determine with high degree of certainty whether persons are biologically related to each other. Determining a relationship with the help of DNA analysis is based on genes inherited from one generation to another. These genes are found in every person's DNA, and half of them are inherited from the biological father and the other half from the biological mother. The father and mother have in turn inherited their genes from their parents.

This explanation not only informs about the principles of genetic kinship but also evokes deep-rooted cultural notions of bilateral relatedness often understood via the idiom of 'blood' in European kinship (Schneider, 1980; see also Carsten, 2011; Franklin, 2013), and therefore, it can make sense without an expert knowledge of genetics. Hence, the accentuated novelty of genetic thinking – or, for that matter, geneticisation – is only relative. When discussing American kinship as a symbolic system, David Schneider (1980: 23) provocatively remarked, 'Kinship is whatever the biogenetic substance is. If science discovers new facts about biogenetic relationship then that is what kinship is and was all along, although it may not have been known at the time'.

Likewise, the symbolism and practices of kinship around the world do contain a plethora of substances, and thus 'social' kinship is indeed biological too (see, e.g.,

Sahlins, 2013). Incidentally, among the Somalis, blood and bones are the vital symbolic substances of kinship, although they are passed on only patrilaterally (see Lewis, 1994). Therefore, the idea of samples – especially when blood samples also are taken – can be comprehensible to Somalis, who have been the ones mostly targeted by DNA analysis in Finland. In addition, the reference to the inheritance of genes between generations also makes cultural sense via the idea of descent.

All in all, the vague reference to 'the material evidence of family ties' in Section 65 (1) of the Aliens Act can and must be invested with cultural meanings. For years, the web pages of the National Institute of Health and Welfare included a simplified chart explaining the method and interpretation of paternity testing. In the chart (see Tapaninen & Helén, 2020: 383), the information about genetic relatedness is put side by side with snapshot-like family portraits by using the symbolism of genealogies: i.e., vertical lines for parentage and horizontal lines for marriage. Interestingly, the people concerned are not represented by the symbols of triangles (male) and circle (female) but by the sketches of the alleged parents and the baby, complete with hairstyles and facial expressions. In the example of confirmed family ties – or, technically, non-exclusion – the parents and the child are happily tied together by the genealogical grid and smiling faces. The second example depicts the negative test results, with the disappointed faces of the mother, alleged father, and even the baby. The exclusion of the man's genetic paternity is highlighted by his exclusion from the genealogical grid too. The picture captivatingly converts genetic markers into genealogical diagrams and finally into images of unified or broken families. Such a translation of 'genetic relatedness between samples' (Anderlik & Rothstein, 2002: 215) into proof of 'true' families has become common knowledge to the extent that it is not seen as a translation at all. While in the decisions of the Finnish Immigration Service, this conceptual leap could be made, this is not always the case, as we will show in the next section: DNA is not enough.

Bureaucratic time: 'Time is on our side'

The Aliens Act does not spell out the scope of DNA testing; it does not specify in which cases DNA analysis could provide 'material evidence of relatedness'. There are no references to the biological basis of parenthood in the definition of family members. Instead, the legal personae include 'guardians' and 'de facto guardians', and family ties are specified by notions of 'custody' and 'care'. This indeterminacy is intensified by the extent of eligibility. While the legislation is relatively inclusive in Finland, it is also ambiguous, and the legal provisions, including the option of DNA analysis, are interpreted in administrative procedures case by case. Therefore, Migri has a vital role in performing the technolegal regime by linking genetics with legislation.

The words of a Migri official define the key issue: 'There must *be* a family to reunify'. Hence, DNA truth is not enough, but time instead is a vital criterion. But how can the existence of a family be verified then? The duality of the 'biological' and 'social' make-up of families is played out in the bureaucratic practices of bordering.

In other words, there have been explicit and consequential dynamics between verification via DNA analysis and more equivocal evidence of 'true' family life.

When the report arrives from the laboratory to Migri for scrutiny, another temporality unfolds. The abstract and timeless make-up of DNA analysis is reversed by punctuated time that retrospectively weaves a connection between the present claims about the family ties and a reconstructed picture of the family life before the sponsor's departure, often by focusing on material details of the applicants' narratives. Even though households in general do not stay the same, and even less so in chaotic situations, a special kind of *Stillleben* is fashioned out of narrative fragments. But this is not enough. In addition, the interviewees must prove that the family ties have not ceased to exist despite separation.

The 254 decisions of the Administrative Court of Helsinki between 2003 and 2013 containing the keyword *DNA* demonstrate the indeterminacy in the quest for 'truth'. How have the evidence of DNA analysis on the one hand and doubting the veracity of family life on the other played out in actual procedures? The appeals attest to the potential of DNA testing for the people concerned. In more than half the appeals, the people demanded testing, which had not been possible for them in the earlier stages of the process.

One-fourth of the appeals clearly show the limits of DNA evidence. In such cases, that evidence had already verified the alleged kinship ties, but the application had been rejected because the other standards of 'genuine' family life were not met (see Halme-Tuomisaari et al., 2019). In our interviews, both the people concerned and their lawyers often found the grounds for rejections absurd. All in all, the demand for permanence is paradoxical because of prolonged separations and precarious circumstances. The temporalities of forced migration complicate the supposedly precise details. The rejections have typically been based on inconsistencies in the details that may derive, according to the appeals, from incomprehension, memory failure or the sheer irrelevance of the questions asked.

The conclusion of the court can be blunt yet non-specific: 'They are not family members in the sense laid down by the Aliens Act'. In rejections, time counts; it has often been argued that family ties ceased to exist when a sponsor left for Finland. Another recurrent basis for a rejection pursuant to Section 36 is that the sponsor had tried to circumvent the provisions by giving false information on the family ties. Importantly, this argument is never based on the result of the DNA analysis but on unconvincing descriptions of intentions and family life.

An immigration official described the challenges faced by the officials in interviews:

> We ask questions about the details of family life – about housing, schools, the everyday life of the family. However, in the Somali context, the children often cannot attend school regularly, the parents may not have jobs, and the quotidian life of the family can be scattered, in a way. It is very difficult to find *markers* with which to piece together an image of family life. And they may all be illiterate with a tenuous conception of time [as compared to ours];

> well, we are sometimes really in trouble, and expertise is called for in evaluating what kind of questions to ask and how to formulate them.

It is not a coincidence that the official chose the word *marker* from the vocabulary of genetics. Rather, it is an indication that the narrative evidence is read in a similar vein, as a reconstruction of snapshot-like minutiae where time is frozen. The shift between the two registers is illustrative of the tendency to translate the criteria of genetic information upon narrative information. From the accounts of the applicants, immigration officials seek precise answers, disregarding the passing of time. For this purpose, as explained in a memorandum, interviews focus on 'detailed and unexpected questions' (Maahanmuuttovirasto, 2013). DNA analysis unavoidably works as a biological lie detector (Lynch et al., 2008; Weiss, 2011; Hall & Naue, 2015), and similarly, interview tactics easily trigger discrepancies. DNA analysis seems to provide the standard for objectivity and precision in decision making, albeit for evidence that is fundamentally different: the fragments of narrative proof (Tapaninen & Helén, 2020). These 'markers' are embedded in the contingencies of time. 'Time is on our side', remarked a police official in our interview with her, and her words capture the predicament of the applicant. Time is definitely not on *their* side.

The grounds for rejections are based not on a definition of acceptable families but rather on the estimation of overall credibility. While the normative make-up of 'true' families directs the verification of acceptable family ties in general (e.g., Carver, 2014; Pellander, 2015), in the case of humanitarian migrants, the perceived strangeness and suspected fraudulence of their relations direct the gaze of immigration control elsewhere. The questions asked often focus on minor material details such as colours of buildings, distance from a water source and dates. Quantifiable and thus disprovable facts seem to count the most. This emphasis on detail is parallel to the logic of DNA analysis. Concomitantly, the credibility of the migrants may be compromised because of 'messy' accounts of family life, which may be related to the maze of the intricate forms of exchange and reciprocity: the very essence of kinship (e.g., Sahlins, 2013).

In 2014, we discussed the conundrums of family reunification with four Finnish Somalis whose applications for family reunification had been rejected. Their cases show that a 'biological' or 'geneticised' definition of family does not necessarily narrow down the applicants' possibilities for family reunification, whereas focusing on the 'social' aspects of family life such as continuity and 'genuineness' in the framework of administrative hermeneutics of suspicion (see Tapaninen & Helén, 2015; Hall & Naue, 2015) often works against their best interest. In fact, 'geneticised' views built on biological evidence might have guaranteed the applicants' right to family reunification.

Our interviewees had escaped from Somalia to Finland in 2007–2008. After that, their families had moved to Addis Ababa or Nairobi, awaiting the slow processing of their applications in the vicinity of the Finnish Embassy. These years witnessed a congestion of applications submitted by Somalis, and thousands of families were stuck. After the tightening of the criteria in 2010, the members of the four families did not manage to join their parent or spouse in Finland. In all cases, rejections were based on minor inconsistencies in interviews. One father told us that two of his nine children, aged ten and eleven, had given indefinite answers regarding former abodes because they had, he explained, been 'fleeing and fleeing'. The experiences of our interlocuters have a lot in common with the cases handled by the Administrative Court. In our interview, they were able to speak out. They said in unison that DNA testing would have been a solution, and they did not understand why it had not been offered. One of them concluded that Migri should frankly say on their web site, 'You will not get your family to Finland'.

Conclusions

In this chapter, we study the use of DNA analysis for family reunification in Finland as a technolegal regime. Looking at this topic from the perspective of technolegal worlds (see Introduction in this volume), family reunification appears as a kind of 'worlding' within which entangled biotechnological, juridical and bureaucratic procedures create specific ontologies for 'the family' as well as criteria for knowing and deciding if a family exists or not and if a person belongs to the family or not (see Tapaninen & Helén, 2020). Our analysis underlines two issues in studying technolegal worlds. First, any such world should be approached as a specific context with sensitivity to political, normative, cultural and even technological nuances (see, e.g., Heinemann et al., 2015). Second, a technolegal world is hardly a unity with a discernible rationale for action; rather, it is an amalgam of epistemic stability and political volatility and is thus characterised by non-coherence and contingency.

Our main purpose in this chapter is to show how evidence on family ties, in the absence of reliable documents, is gathered and interpreted by the Finnish Immigration Service (Migri) via interviews and DNA analysis. We analyse the complex and rather equivocal relationship of the two forms of evidence, initially to challenge the commonplace notions based on the dualism of 'biological' (or genetic) and 'social' families. We find the critical views towards the 'geneticization' of families via biotechnologies (e.g., Heinemann & Lemke, 2015) too narrow and reductive a conceptualisation.

To bypass this dualism, we approach family reunification as a technolegal regime (see Introduction in this volume), in which DNA analysis and interview narratives are used and interpreted in multiple legal and administrative settings. Our analysis shows that within this technolegal regime – including repeatedly amended legislation, a variety of administrative instances, the laboratory and the appellant courts – relatedness as a 'family' is complex and equivocal, and it can be made, remade and unmade (Tapaninen & Helén, 2020) over time.

In the investigations by immigration authorities, family ties are to be verified – and thus defined, modified and constructed (see Tapaninen & Helén, 2020) – under three temporal orders. *The time of law* entails frequent amendments of the Aliens Act that alter the weight of DNA testing, even though the provisions have been largely unaltered for two decades. In the legislation itself, the temporal indications are mostly implicit; however, their significance has been modified due to administrative interpretations that make time a critical issue for the applicants. The criterion of continuity in the definition of 'genuine' family life is particularly relevant.

DNA analysis brings *technoscientific time* into the family reunification procedures, but paradoxically, it sets temporality aside almost completely. Technically, it only establishes whether the alleged genetic relatedness between the two samples given by two persons exists or not. Thus, it provides a snapshot of biological relatedness that is translated by the immigration officials into an equally timeless picture of family relations. In Finland, however, the result of DNA analysis is not decisive when immigration authorities make a decision. What matters most is the evidence of the 'permanence 'and 'genuineness' of family life that the officials try to extract from interviews of applicants. However, the search for 'good markers' via interviews also mirrors the logic of DNA analysis. In immigration officials' decisions, technoscientific time is subsumed by *bureaucratic time*.

The applicants' accounts – or rather answers to the questions posed – in the interviews are recollections detailing myriad minutiae of their family life before the sponsor's departure and during their separation. Immigration officials try to verify the existence of 'true' families on the basis of these accounts, and thus, the administrative procedure implies a temporal order that refers to the past and is constituted by recollections presented in the specific context of an immigration interview (on the latter, see Puumala et al., 2018). These recollections are fractured for two reasons. First, paradoxically, the lived worlds depicted cannot be stable enough because of separations and circumstances that are volatile and insecure. Second, the questions asked in interviews focus on minor, perhaps incomprehensible or irrelevant 'markers', which makes the descriptions even more fractured and potentially inconsistent. The time of fractured recollections of a scattered family is a basic element of bureaucratic time. The bureaucratic time of control unfolds through two largely contrasting temporalities: the demand for permanence and the questions digging up minutiae that are out of time, especially from the time of practiced kinship. In fact, we argue that the complexities of kinship are largely overlooked or negated in the proceedings.

These tensions inherent in the bureaucratic time show that there hardly exists a unified time order in the technolegal regime of the Finnish family reunification. Rather, speaking of 'technolegal time' in this setting refers to mutually noncoherent or even contradictory temporalities that remain quite implicit. This reflects the equivocality, contingency or even arbitrariness that characterises immigration management in general.

These observations are not all there is to family reunification, however, because the temporalities of immigration management are inevitably entangled with the contingencies of the time of migration. The journey of a refugee is volatile and takes a lot of time. In addition, the process by which a person is allowed or disallowed a residence permit or family reunification takes time, as administrative and juridical phases of the procedures have their own logic and duration. This is an

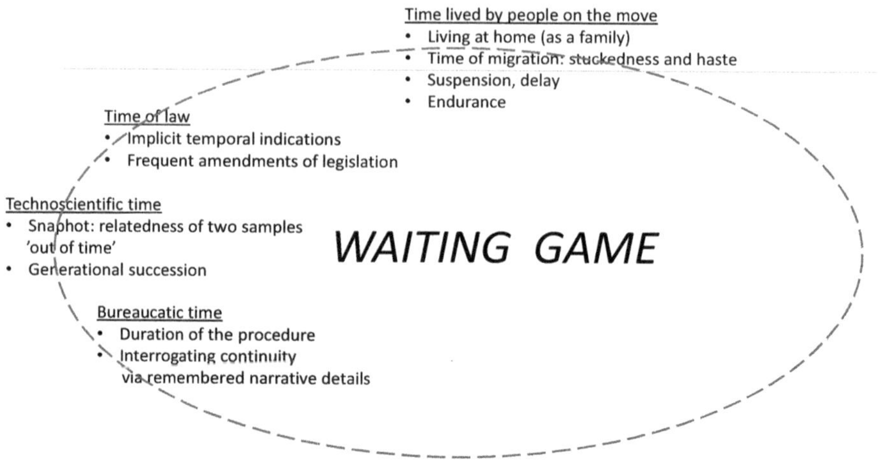

FIGURE 10.1 The temporalities of family reunification

important element of the temporal ordering of immigration bureaucracy: procedures such as family reunification endure, which certainly influences the ways officials assess and decide on such cases. For the people on the move trying to join their kin, this aspect of bureaucratic time appears as a delay or suspension. Time in the process is manifested and imposed as immobility and stuckedness, which characterises the condition and experiences of migrants more generally. In many ways, family reunification is also a waiting game (see Könönen, 2019), in which time stands still. Immigrants try to fight obstacles – and time – with the best information and resources at hand, but the requirements and interpretations keep changing, often in unforeseeable ways.

Notes

1 We certainly agree with Haimes and Toom (2014; see also Toom, 2016) that kinship should be taken more thoroughly into account in the proceedings of and research on victim identification. The same applies to family reunification, but because of the complexity of kinship, here we can discuss it only parenthetically.

2 The first project was *DNA and immigration: Social, political and ethical implications of DNA analysis for family reunification* (IMMIGENE), funded by the Academy of Finland (grant 135266) and comparing Austria, Finland and Germany (see Heinemann et al., 2015), and the second was *Bodies of evidence: The interplay of documents, narratives, and biotechnologies in immigration control*, funded by the Kone Foundation. In addition, the data was updated in an ongoing research project called *Struggles over home and citizenship: Neighbourhood solidarity as a response to the 'Asylum Crisis'*, also funded by the Kone Foundation.

3 In the context of family reunification in the EU, the term *sponsor* refers to a 'third-country national' with a residence permit whose family members apply for family reunification to join her or him. Here, we use this legal term because it makes the logic clear enough: the options are largely defined by the sponsor's formal status.

4 At that time, the rising numbers of asylum seekers and, concomitantly, of applications for family reunification became a heated political issue. The backlog of applications filed by Somalis was regularly reported on by Migri and pointed out by the media. At the time, the applicants lived through prolonged stuckedness.
5 The only precise definition is the maximum time of nine months set in Section 69a for the processing of family reunification applications. Especially at the turn of the 2000s, this was often exceeded. Even though, pursuant to Section 6 (3), 'matters concerning minors shall be processed with urgency', after the restrictions, the process was exceptionally long for the guardians of children under international protection: in 2016, the process took on average 325 days, and most of the applications were rejected.
6 In 2019, one-fifth of the sponsors in family reunification cases were recipients of international protection (EMN, 2019). Even though DNA testing does not concern only people with a refugee background, they are an overwhelming majority of those tested. Hence, in relation to the number of laboratory tests, around 80% of asylum seekers' families were tested – 1,379 tested persons from 440 families.
7 The reliability of DNA analysis is based on data on the frequency of genetic markers among the relevant population, however it is defined (see M'charek, 2000). In Finland, such information has been collected on Somalis (see Neuvonen et al., 2012). Finnish geneticists point out, however, that the information does not add up to an actual database.
8 Since 2016, all DNA analyses have been centralised in the laboratory of the National Institute of Health and Welfare. Before that, most analyses were made by the Department of Forensic Medicine at the University of Helsinki.
9 We are not suggesting that forensic genetics is timeless (see M'charek, 2014). Instead, we want to point out that in the context of DNA testing for family reunification, laboratory analyses, in contrast to other proof and the time of migration, are out of time.

References

Anderlik, M. & Rothstein, M.A. 2002. DNA-based identity testing and the future of the family: A research agenda. *American Journal of Law and Medicine* 28:2–3, 215–232.

Andersson, R. 2014. Time and the migrant other: European border controls and the temporal economy of illegality. *American Anthropologist* 116:4, 795–809.

Bourdieu, P. 1977. *Outline of a Theory of Practice*. Cambridge: Cambridge University Press.

Browner, C.H. and Sargent, C.F. 2011. *Reproduction, Globalization, and the State: New Theoretical and Ethnographic Perspectives*. Durham, NC: Duke University Press.

Cabot, H. 2012. The governance of things: Documenting limbo in the Greek asylum procedure. *PoLAR: Political and Legal Anthropology Review* 35:1, 11–29.

Carsten, J. 2011. Substance and relationality: Blood in contexts. *Annual Review of Anthropology* 40, 19–35.

Carsten, J. 2013. What kinship does—and how. *HAU: Journal of Ethnographic Theory* 3:2, 245–251.

Carver, N. 2014. Displaying genuineness: Cultural translation in the drafting of marriage narratives for immigration applications and appeals. *Families, Relationships and Societies* 3:2, 271–286.

Dove, E.S. 2013. Back to blood: The sociopolitics and law of compulsory DNA testing of refugees. *University of Massachusetts Law Review* 8, 466–530.

EMN. 2016. EMN Ad-Hoc query on checking identity and family relationships in case of family reunification with a beneficiary of international protection. Available at: https://ec.europa.eu/home-affairs/sites/homeaffairs/files/what-we-do/networks/european_migration_network/reports/docs/ad-hoc-queries/ad-hoc-queries-2016.1074_nl_checking_identity_and_family_relationships.pdf.

EMN. 2017. EMN synthesis report for the EMN focussed study 2016. Family reunification of third-country nationals in the EU plus Norway: National practices. Available at: https://ec.europa.eu/home-affairs/sites/homeaffairs/files/00_family_reunification_synthesis_report_final_en_print_ready_0.pdf.

EMN. 2019. Annual report on migration and asylum. Finland. Available at: www.emn.fi/files/1970/Maahanmuutto_ja_turvapaikka_2018_FI_EN.pdf

Eurostat. 2019. Residence permits—statistics on first permits issued during the year. Available at: https://ec.europa.eu/eurostat/stati teststics-explained/index.php?title=Residence_permits_-_statistics_on_first_permits_issued_during_the_year&oldid=456573.

Fassin, D. & Halluin, D. 2005. Truth from the body: Medical certificates as ultimate evidence for asylum seekers. *American Anthropologist* 107:4, 597–608.

Feldman, G. 2011. If ethnography is more than participant-observation, then relations are more than connections: The case for nonlocal ethnography in a world of apparatuses. *Anthropological Theory* 11:4, 375–395.

Finnish Immigration Service. 2008. DNA analysis has unified a record number of immigrant families. Available at: www.migri.fi/for_the_media/bulletins/press_releases/press_releases/1/0/dna_testing_has_united_a_record_number_of_immigrant_families_13968.

Finnish Immigration Service. 2019. DNA testing in connection with a residence permit application. Available at: https://migri.fi/documents/5202425/6160286/DNA+testing+in+connection+with+a+residence+permit+application+%28en%29.pdf/f0526cac-4e86-41a7-a1e5-f2a72ec8c3cf/DNA+testing+in+connection+with+a+residence+permit+application+%28en%29.pdf?version=1.5&t=1569583373000.

Franklin, S. 2013. From blood to genes? Rethinking cosanguinity in the context of geneticization. In *Blood and Kinship: Matter for Metaphor from Ancient Rome to the Present*, edited by C. H. Johnson, B, Jussen, D. W. Sabean & S. Teuscher. New York and Oxford: Berghahn, 285–320.

Ginsburg, F. & Rapp, R. (eds.) 1995. *Conceiving the New World Order: The Global Politics of Reproduction*. Oakland, CA: University of California Press.

Good, A., Berti, D. & Tarabout. G. 2015. Introduction: Technologies of doubt in law and ritual. In *Of Doubt and Proof: Ritual and Legal Practices of Judgment*, edited by D. Berti, A. Good & G. Tarabout. Farnham: Ashgate, 1–18.

Granados Moreno, P., Ngueng Feze, I. & Joly, Y. 2017. Does the end justify the means? A comparative study of the use of DNA testing in the context of family reunification. *Journal of Law and the Biosciences* 4:2, 250–281.

Griffiths, M.B. 2014. Out of time: The temporal uncertainties of refused asylum seekers and immigration detainees. *Journal of Ethnic and Migration Studies* 40:12, 1991–2009.

Hage, G. 2009. Waiting out the crisis: On stuckedness and governmentality. In *Waiting*, edited by G. Hage. Carlton, VIC: Melbourne University Press, 97–106.

Haimes, E. & Toom, V. 2014. Hidden in full sight: Kinship, science and the law in the aftermath of the Srebrenica genocide. *New Genetics and Society* 33:3, 277–294.

Hall, K. & Naue, U. 2015. Austria: DNA profiling as a lie detector. In *Suspect Families*, edited by T. Heinemann et al. Farnham: Ashgate, 55–78.

Halme-Tuomisaari, M., Tapaninen, A-M. & Aunela, H. 2019. Where's the well: DNA evidence, personal narratives and unpredictability in Finnish family reunification. *Migration Studies* 7:2, 245–265.

Harbers, H. 2005. How much time can we stand? DNA evidence and the principle of finality in criminal law. *Configurations* 13, 357–371.

HE 240. 2009. Hallituksen esitys eduskunnalle laiksi ulkomaalaislain muuttamisesta. Available at: www.eduskunta.fi/FI/vaski/HallituksenEsitys/Documents/he_240 + 2009.pdf.

Heinemann, T. & Lemke, T. 2013. Suspect families: DNA kinship testing in German immigration policy. *Sociology* 47:4, 810–826.

Heinemann, T. & Lemke, T. 2015. Germany: The geneticisation of the family. In *Suspect Families*, edited by T. Heinemann et al. Farnham: Ashgate, 13–32.

Heinemann, T., Helén, I., Lemke, T., Naue, U. & Weiss, M. (eds) 2015. *Suspect Families: DNA Analysis, Family Reunification and Immigration Policies*. Farnham: Ashgate.

Heinemann, T., Naue, U. & Tapaninen, A-M. 2013. Verifying the family? A comparison of DNA analysis for family reunification in three European countries (Austria, Finland and Germany). *European Journal of Migration and Law* 15:2, 183–202.

Helén, I. & Tapaninen, A.M. 2013. Closer to the truth: DNA profiling for family reunification and the rationales of immigration policy in Finland. *Nordic Journal of Migration Research* 3:3, 153–161.

Holland, E. 2011. Moving the virtual border to the cellular level: Mandatory DNA testing and the U.S. refugee family reunification program. *California Law Review* 99:6, 635–681.

Jacobsen, C.M., Karlsen, M.A. & Khosravi, S. (eds.) 2020. *Waiting and the Temporalities of Irregular Migration*. London & New York: Routledge.

Jeffreys, A.J., Brookfield, J.F. & Semeonoff, R. 1985. Positive identification of an immigration test-case using human DNA fingerprints. *Nature* 317:6040, 818–819.

Kelly, T. 2015. Afterword. In *Of Doubt and Proof: Ritual and Legal Practices of Judgment*, edited by D. Berti, A. Good & G. Tarabout. Farnham: Ashgate, 183–193.

Khosravi, S. 2010. *'Illegal' Traveller: An Auto-ethnography of Borders*. Berlin: Springer.

Könönen, J. 2019. The waiting game: Immigration detention as the waiting room of immigration law. *Migration Studies*. AOP DOI: 10.1093/migration/mnz036.

Kruse, C. 2015. *The Social Life of Forensic Evidence*. Berkeley: University of California Press.

La Spina, E. 2012. DNA testing for family reunification in Europe: An exceptional resource? *Migraciones Internacionales* 6:22, 40–75.

Lee, C. & Voigt, T.H. 2020. DNA testing for family reunification and the limits of biological truth. *Science, Technology & Human Values*. AOP DOI: 10.1177/0162243919862870.

Leinonen, J. & Pellander, S. 2020. Temporality and everyday (in)security in the lives of separated refugee families. In *Family Life in Transition: Borders, Transnational Mobility, and Welfare Society in Nordic Countries*, edited by J. Hiitola, K. Turtiainen, S. Gruber & M. Tiilikainen. London & New York: Routledge, 118–128.

Lems, A. & Tošić, J. 2019. Preface: Stuck in motion? *Journal of the Finnish Anthropological Society* 44:2, 3–19.

Lewis, I.M. 1994. *Blood and Bone: The Call of Kinship in Somali Society*. Trenton, NJ: The Red Sea Press.

Lynch, M., Cole, S.A. & McNally, R. 2008. *Truth Machine: The Contentious History of DNA Fingerprinting*. Chicago: University of Chicago Press.

Maahanmuuttovirasto. 2013. *Oleskelupa perhesiteen perusteella*. Unpublished memorandum 13.5.2013. Helsinki: Maahanmuuttovirasto.

M'charek, A. 2000. Technologies of population: Forensic DNA testing practices and the making of differences and similarities. *Configurations* 8, 121–158.

M'charek, A. 2014. Race, time and folded objects: The HeLa error. *Theory, Culture & Society* 31:6, 29–56.

M'charek, A. 2020. Harraga: Burning borders, navigating colonialism. *The Sociological Review* 68:2, 418–434.

M'charek, A., Toom, V. & Jong, L. 2020. The trouble with race in forensic identification. *Science, Technology, & Human Values*. AOP DOI: 10.1177/0162243919899467.

McKinnon, S. & Cannell, F. (eds.) 2013. *Vital Relations: Modernity and the Persistent Life of Kinship*. Santa Fe, NM: SAR Press.

Mol, A. 2002. *The Body Multiple: Ontology in Medical Practice*. Durham, NC & London: Duke University Press.

Näre, L. 2020. Family lives on hold: Bureaucratic bordering in male refugees' struggle for transnational care. *Journal of Family Research* 32:3, 435–454.

Netz, S. 2019. Enacting biorelational citizenship in X-ray rooms and offices: On age assessments of migrants in Germany. *Citizenship Studies* 23:7, 720–737.

Netz, S., Lempp, S., Krause, K. & Schramm, K. 2019. Claiming citizenship rights through the body multiple. *Citizenship Studies* 23:7, 637–651.

Neuvonen, A.M., Palo, J.U., Hedman, M. & Sajantila, A. 2012. Discrimination power of Investigator DIPplex loci in Finnish and Somali populations. *Forensic Science International: Genetics* 6:4, e99–e102.

Olwig, K.F. 2020. The right to a family life and the biometric 'truth' of family reunification: Somali refugees in Denmark. *Ethnos*. AOP DOI: 10.1080/00141844.2019.1648533.

Pellander, S. 2015. 'An acceptable marriage': Marriage migration and moral gatekeeping in Finland. *Journal of Family Issues* 36:11, 1472–1489.

Puumala, E., Ylikomi, R. & Ristimäki, H-L. 2018. Giving an account of persecution: The dynamic formation of asylum narratives. *Journal of Refugee Studies* 31:2, 197-215.

Rozakou, K. 2020. The violence of accelerated time. In *Waiting and the Temporalities of Irregular Migration*, edited C.M. Jacobsen, M-A. Karlsen & S. Khosravi. London & New York: Routledge, 23–39.

Sahlins, M. 2013. *What Kinship Is—and Is Not*. Chicago: University of Chicago Press.

Schneider, D. 1980. *American Kinship: A Cultural Account*. Chicago: University of Chicago Press.

Taitz, J., Weekers, J.E.M. & Mosca, D.T. 2002. The last resort: Exploring the use of DNA testing for family reunification. *Health and Human Rights* 6:1, 20–32.

Tapaninen, A.M. 2018. Rajalla laboratoriossa: iän arviointia, iän määrittämistä ja ikämääräyksiä. *Tiede ja edistys* 43:3, 216–237.

Tapaninen, A-M., Halme-Tuomisaari, M. & Kankaanpää, V. 2019. Mobile lives, immutable facts: Family reunification of children in Finland. *Journal of Ethnic and Migration Studies* 45:5, 825–841.

Tapaninen, A.M. & Helén, I. 2015. Finland: Securing human rights, suspecting fraud. In *Suspect families*, edited by T. Heinemann et al. Farnham: Ashgate, 33–53.

Tapaninen, A.M. & Helén, I. 2020. Making up families: How DNA analysis does/does not verify relatedness in family reunification in Finland. *BioSocieties* 15:3, 376–393.

Ticktin, M. 2011. How biology travels: A humanitarian trip. *Body & Society* 17:2–3, 139–158.

Toom, V. 2016. Whose body is it? Technolegal materialization of victims' bodies and remains after the World Trade Center terrorist attacks. *Science, Technology, & Human Values* 41:4, 686–708.

van der Ploeg, I. 1999. The illegal body: 'Eurodac' and the politics of biometric identification. *Ethics and Information Technology* 1:4, 295–302.

Villiers, J.D. 2010. Brave new world: The use and potential misuse of DNA technology in immigration law. *Boston College Third World Law Journal* 30:2, 239–271.

Weiss, M.G. 2011. Strange DNA: The rise of DNA analysis for family reunification and its ethical implications. *Genomics, Society and Politics* 7:1–19.

11

EVALUATING FORENSIC DNA DATABASES

Carole McCartney and Aaron Opoku Amankwaa

11.1 Introduction

In March 2018, Andrew Pennington received a ten-year prison sentence for rape and three burglaries committed in 1988 and 1997, respectively. The police made a breakthrough during a cold case review, when a semen stain on the skirt of the victim yielded a DNA profile which matched a reference DNA profile of Pennington held on the UK National DNA Database (UK NDNAD) (North Yorkshire Police, 2018). The Pennington case neatly illustrates the multiple roles that a forensic DNA database can play in criminal investigations. A match from the database can lead to the apprehension of previously unsuspected offenders (a so-called 'cold hit') while eliminating individuals from a criminal inquiry. (The original suspect here was belatedly excluded.) The database can also link different crimes (rape and burglaries here) to identify serial offenders (with such links also supporting criminological research into crime patterns; see Struyf et al., 2019). Databases can also be searched in efforts to solve international/transnational offending with cross-border law enforcement efforts (particularly in the EU with the semi-automated Prum system).

The Pennington case also serves as an archetype for stories of serious criminals caught only by virtue of the DNA database, eulogised by advocates of DNA databasing. Prainsack and Toom (2013) explain how certain cases become 'founding myths', creating a narrative with DNA technology lauded as 'triumphant hero' (Aronson, 2007), securing significant financial investment and political commitment to expanding forensic DNA databases. It might then be presumed from such investment, and the prevalence of such stories and the significance accorded to DNA 'matches', that DNA databases make a vital contribution to criminal justice aims. Indeed, examples of such claims abound. In 2009, the UK pronounced it was leading the world in the use of DNA 'to solve crimes, to catch criminals and to

DOI: 10.4324/9780429322358-15

clear the innocent. Without this ability, we would be less safe, and criminals would be more likely to get away with their crimes' (Home Office, 2009, para. 2.1). A cursory scan of countries that have established national forensic DNA databases reveals similar assertions. An Australian police minister asserted, 'We know DNA is a more effective way for police to prosecute and solve crime', arguing that DNA databases are 'helping police solve thousands of unsolved crimes and helping to catch serious offenders more easily' (Pearson, 2018). A Russian interior ministry spokesperson, meanwhile, has declared that DNA tests were 'one of the most effective instruments in criminal investigation', with the Russian federal DNA database identifying over 70,000 perpetrators: 'most of them serious and particularly serious' (Tass, 2021). Headlines in India, meanwhile, claim 'Offender DNA Database Will Be a Strong Deterrent to Repeat Crimes in India', citing 'prominent international studies' proving that it will contribute 'more to public safety than conventional punitive measures' and that it is 'imperative' that India have a DNA database (Business Wire India, 2021).

Such claims are echoed in pronouncements that the Prum regime of DNA exchange across the EU has been 'very useful', an inference that remains largely unsubstantiated (Toom et al., 2019). Such contentions then prompt the question of how it is known that DNA databases are 'more effective', solving crimes 'more easily', making it 'imperative' that a state has one, when the authority and reliability of such claims remain obscure at best. Rudimentary statistics (such as 'cases solved', 'perpetrators identified') are often cited, and yet, 'the significance of what can be gleaned from such data, particularly the limitations, is not widely understood' (IAG, 2018). Proponents of DNA databasing have often been 'generous' in their representations of the success of DNA profiling since the inception of DNA databasing. The Human Genetics Commission (2009:7) bemoaned the 'widespread use of exaggerated rhetoric, which replaces, or obscures reasoned argument and the careful examination of the evidence'. Indeed, misrepresentations and misunderstanding of DNA databases and their 'performance' persist, with prominent publication of inaccurate statements, such as '[T]he database [UK NDNAD] helps to identify a suspect in around 60% of criminal cases' (Black, 2019).

High-profile crimes for which DNA had been useful during an investigation are thus given significant publicity, although many cases for which DNA could have been expected to, but did not, assist police investigations go unmentioned, just as those cases in which DNA evidence led to 'tunnel vision' (a recognised precursor to miscarriages of justice) or a wild goose chase are brushed aside.[1] Failure to use the DNA technology and police powers available has also prolonged police investigations (with growing victim tallies), such as that of the multiple rapist and killer Robert Napper (Fraser, 2020). Indeed, when promoting the UK NDNAD during its formative years, 'Failures of the science, or the police to utilise the science, no longer fitted the official story of the 'new dawn' being witnessed in policing' (McCartney, 2017a:418), a story that continues to be told today.

While the value of a DNA database in individual (particularly unsolved) cases like Pennington's is relatively easy to demonstrate (although investigations are not normally that straightforward, and detections can rarely be solely attributed to a DNA 'match'), the aggregate value of DNA databases remains unascertained or exaggerated (Amankwaa and McCartney, 2021) It is also often conflated with the value of DNA evidence more generally, or even 'forensic evidence' as a whole, rather than specifically the use of a database. Various measures have been proposed to evidence the effectiveness of DNA databases, each typically reflecting very narrow parameters of assessment. Attempts to evaluate databases are also necessarily dependent on prior expectations, which mostly remain ill-defined and ordinarily omit any complementary account of broader impacts, which are similarly often vague and disputed (Amankwaa and McCartney, 2019). Measures of 'effectiveness' must be dissected so that they are more realistic, and evaluations must also be expanded to incorporate broader but integral requirements – validity, legitimacy and acceptability – to ensure the 'integrity' of DNA databasing.

An effective justice system 'must assess itself not only against narrow criteria of crime control, but against broader criteria relating to people's trust in justice and their sense of security' (European Commission, 2011:11). Integrity is essential to maintaining trust in judicial bodies, law enforcement agencies and justice systems more broadly, trust being necessary to secure widespread commitment to the rule of law and normative compliance with the social order (Hough and Sato, 2011). 'Integrity' is a multi-faceted but easily understood concept that can refer to both physical condition (e.g., being 'whole' or 'stable') and moral and ethical standing (e.g., being 'honourable' or 'principled'). References to 'integrity' in English and Welsh courtrooms as well as in Scottish, Irish and Commonwealth courts proliferate in a variety of contexts, being invoked in reference to, *inter alia*, witnesses, policing, evidence, juries, trials, the administration of justice, the criminal process and the criminal justice system as a whole (Hunter *et al.*, 2016:7). Integrity thus offers 'a powerful conceptual lens through which the criminal process in its entirety, or selected phases or aspects of it, can be viewed and critically re-examined' (Hunter *et al.*, 2016:1). The interrelationship between integrity, confidence, justice and security, and the significant role these have in ensuring trust both within and outside justice systems are belatedly coming into focus with respect to technologies being introduced to criminal justice and other 'security' systems.

The 'integrity' of a DNA database should be assessed against a matrix of standards to be met, which, when combined, would provide a holistic evaluation of a DNA database. To ensure the integrity of forensic DNA databases, essential for generalised trust among not just the stakeholders of DNA databases but the public also, critical attention must be paid not only to the viability of forensic DNA databases but also to their legitimacy and acceptability (McCartney, 2017b:170), including the ethical erasure of material/records (Skinner and Wienroth, 2019). This chapter will thus first examine discourses around the effectiveness of DNA databases (with a focus on the UK NDNAD) before sketching an approach to evaluation that could augment anecdotal evidence and elementary statistical analysis, inaugurating a more realistic and holistic account of DNA databasing, focusing on the 'integrity' of databasing. Section 11.2 examines the 'public security' outcomes by which the effectiveness of DNA databases has been measured to date, before discussing in 11.3 how these effectiveness criteria need to be 'dissected' to enable a more realistic evaluation. Section 11.4 then suggests the parameters of a more holistic evaluation of the effectiveness of criminal forensic DNA databases.

11.2 Measuring effectiveness: Public security aims

To determine the effectiveness of forensic DNA databases, a measure must be made of the success of such databases in achieving their aims (Amankwaa and McCartney, 2019). An important interrelated demand is that the database is also 'efficient': i.e., the 'cost' of the achievement of those aims is favourably compared to that of alternative systems (a cost/input-benefit analysis) (Amankwaa and McCartney, 2019). Of course, both measures are dependent on the aims of the database. In the UK, the Police and Criminal Evidence Act (PACE, 1984) stipulates the justifications for taking and retaining DNA samples: 1) in the interests of national security; 2) for the purposes of a terrorist investigation; 3) for purposes related to the prevention or detection of crime, the investigation of an offence or the conduct of a prosecution; or 4) for purposes related to the identification of a deceased person or of the person to whom the material relates (PACE Section 63T(1)). Most countries will have similar stated aims in enabling legislation, although some will have drawn narrower objectives (focusing perhaps on serious crime) while some may be more indeterminate. For example, in Australia, Queensland's legislation permits the police commissioner to use their DNA database for *any* police service function (Queensland Police Powers and Responsibilities Act, 2000: Section 493).

In the UK, an effective database should then contain both relevant subject reference profiles and crime scene profiles that together generate 'matches' that enable the police and prosecutors to prevent, detect, investigate and prosecute crime and terrorism and identify unknown decedents, which could conjointly be considered as enhancing 'public security' (Amankwaa and McCartney, 2019). An efficient forensic DNA database system should result in better public security outcomes than the use of alternative tools. At the very least, the public security outcomes should merit the input required (the 'costs') of the operation of the DNA database (i.e., it is not significantly less efficient than other policing methods). However, there is severely limited data to demonstrate the effectiveness of the NDNAD in achieving these public security goals. The biometrics commissioner refers in part to the problem of multiple variables as one reason for this gap, commenting that:

> a knowledge base on the effectiveness of the use of both DNA and fingerprints in police investigations does not exist, in part because it is very difficult to identify the added value from biometrics compared to other information available during an investigation.
>
> *(Wiles, 2019)*

Scholars have taken differing approaches to measuring the effectiveness of DNA databases. Bieber (2006) recommended three outcomes. The first is case resolution effect (CRE), which focuses on how cold hits contribute to the successful resolution of criminal investigations, including the database conviction rate, timeliness (speeding up detections) and reducing the cost of investigations. The second outcome is a crime prevention effect (CPE), which examines how DNA databases assist in efforts to prevent future crime through the incapacitation of offenders or deterrence effects. Bieber also briefly considers a 'societal and individual interests' (SII) measure, covering perceptions of public security benefits, privacy and related civil liberties. Krimsky and Simoncelli (2011) proposed a mathematical equation to determine what they call the 'Crime Solving Efficiency – CSE'. However, their 'CSE index' only considers convictions as the outcome of databases, yet convictions are not an accurate measure as there are too many confounding variables, and it is difficult (if not impossible) to determine whether a DNA match was the catalyst for a conviction since it is most often considered alongside other corroborative evidence (Bieber, 2006; Goulka *et al.*, 2010). Indeed, it is this reasoning that is most frequently relied on by authorities to justify a lack of data on DNA convictions (FIND Strategy Board, 2020:20, 26). Therefore, any proposed CSE calculations will always flounder because of a lack of data and the necessarily attendant legion caveats.

A different approach using database output variables, termed 'match rates', is proposed by van der Beek (2015), relying on a mathematical model provided by Walsh et al. (2010) utilising two efficiency parameters: H/C and H/N, where H/C measures the chance of a crime stain profile matching a reference profile in the database, and H/N measures the fraction of reference profiles loaded in the database that have contributed to a match to a crime stain profile. Whilst this approach is currently utilised by several national DNA databases, including the UK NDNAD, the match rates do not represent actual effectiveness or efficiency because DNA matches require confirmation of whether the matched subject is a suspect (rather than victim/witness etc.). Also, the DNA match may not have been relevant to the resolution of the case; in many cases, identity is not in question, or the suspect was also identified by other means.

Taking a broader view, key indicators could include not only the crime-solving capacity of the database but also any incapacitation and deterrence effects and also consider privacy protections and the proportionality of the database, alongside implementation 'costs' and efficiency comparators (Amankwaa and McCartney, 2019). While offering a more comprehensive account of the impact of a database, such indicators would pose significant problems with respect to obtaining relevant data. Presently, oversight bodies rely heavily on the primary metric in reports: match rates. Attempts to move beyond bald match rates, with data for the number of investigations in which DNA had been 'linked to outcome', are limited by the fact that any 'link' and its importance in an investigation are left unexplained.

The results are also seriously underwhelming: DNA was 'linked to outcome' in just 0.3% of all recorded crimes in England and Wales in 2015–2016 (Wiles, 2017).

The highest rate was recorded for homicides (8.4%), followed by domestic burglaries (1.4%), theft of vehicles (0.9%) and rapes (0.6%). The 2018–2020 NDNAD report simply states under 'Outcomes', 'The number of offenders convicted with the help of DNA evidence is not recorded', followed by three case 'vignettes' in which DNA database matches helped detect an offender. Reverting to reliance on individual cases (again) to demonstrate (positive) impact, with muted concession that, with respect to aggregate 'outcomes', these are unknowable, and attempts are not made to report on them. This data, of course, should also be read in the broader context of a falling rate in the number of all crimes that are 'detected', which has dropped to historically low levels, with just 7.5% of recorded crimes in England and Wales in 2020 leading to a suspect being charged or ordered to appear in court. In fact, in 38.5% of recorded crimes in 2020, no suspect was identified (Home Office, 2021). Whatever the confounding variables that may be impacting the ability of the DNA database to improve detection rates, the available evidence indicates the aggregate contribution of the NDNAD to the resolution of crime remains stubbornly low.

11.3 Dissecting effectiveness

These 'public security' parameters of effectiveness are thus incredibly narrow, and 'effectiveness' needs dissecting, with a wider range of criteria to provide a more realistic and holistic accounting of DNA database impact. Amankwaa (2019) conceptualised database effectiveness as comprising of four main elements: actual, potential, perceived and cumulative effectiveness. 'Actual' effectiveness, as seen in the studies mentioned earlier, evaluates the currently measured outcomes of a system and how it meets standardised outcomes, such as 'match rates'. Potential effectiveness goes further than these easily counted, elementary measures, with a further assessment of the capacity of a system to meet its stated purpose(s). Perceived effectiveness concerns the attitudes or beliefs of citizens about whether a system is achieving its purpose or not. Cumulative effectiveness, then, would be a combination of these aspects of effectiveness and may also incorporate 'expanded' measures of evaluation.

11.3.1 'Actual' effectiveness

Actual effectiveness investigates the contribution of the database to the specified public security outcomes. As discussed earlier, realistically, this parameter is impossible to assess. However, studies and reports have focused on 'matches' and extrapolate from 'hits' evidence of impact on crime resolution and even deterrence. The most recent systematic review of studies using 'matches' as a measure of effectiveness indicates some positive evidence for both their case resolution and their preventative effects (Struyf *et al.*, 2019). However, as we have seen, this is highly variable and dependent on the crime type. Further, there is no data on how the contribution of database matches compares with other evidence types, such as witness statements, fingerprint evidence and direct comparison of DNA profiles without the use of a database.

Bramley (2009) suggested that the 'crime prevention rate' due to the incapacitation effect of the UK NDNAD was 7.8 crimes per custodial sentence. These values have been criticised for a lack of clarity on the methods used to arrive at the estimates (McCartney, 2006), as well as the fact that any 'incapacitation effect' will be severely limited by the small minority of offenders who are given custodial sentences. Still, the 'deterrent effect' of DNA databases is regularly invoked: that individuals will desist from crime since the database increases their chances of being caught. Further, an 'incapacitation' effect means that offenders caught by DNA databases are unable to commit further crimes while imprisoned (Doleac, 2017). In the US (Doleac, 2017) and Denmark (Anker *et al.*, 2021), investigations into a 'return on investment' of a DNA database find that while initial costs are high, the cost of adding profiles to an established database become marginal, so only small decreases in crime are needed to 'justify' the financial costs of continued expansion. However, there are several confounders with the association of DNA databasing to crime rates and recidivism while opportunistic or impulsive criminals do not tend to consider their chances of detection. Recidivistic offenders may also change their *modus operandi* to prevent detection rather than desisting from crime (McCartney, 2006). Considering these limitations, 'proof that any single new program in the justice system directly reduces crime rates would be difficult to convincingly demonstrate statistically' (Bieber, 2006:230). Any deterrence or incapacitation effect is yet to be convincingly demonstrated, along with any impact on crimes that do not involve DNA and the increasing number that are never investigated by the police.

11.3.2 Potential effectiveness

While the potential effectiveness of the database would seem to be enhanced by having as many DNA profiles on the database as possible, this is not true. After the deletion of over 1 million profiles following the ruling that the UK DNAD breached the European Convention on Human Rights in *S and Marper v the UK* [2008] ECHR 1581, the NDNAD match rate increased. It is not the size of the database that determines its effectiveness, but other factors are critical, such as whose data is retained and whether the NDNAD is used to best effect with respect to loading DNA profiles and the police follow-up of DNA 'matches' (see Amankwaa and McCartney, 2021). Of course, DNA databasing cannot be held to blame for the inefficiencies of other systemic reasons for low crime detection and prosecution rates. Previously, the seemingly intractable inefficiencies of the processes preceding and following utilisation of the NDNAD (effective crime scene examinations and investigations following a match) have been laid bare, with DNA matches falling into 'black holes' (HMIC, 2012) and DNA evidence referred to as 'a fresh filling between two slices of stale bread' (Leary and Pease, 2003:11).

To extricate the NDNAD from this disappointing sandwich, a measure of potential effectiveness should serve to demonstrate the impact that a DNA database could have on public security aims, using the H/C and H/N match rates. The crime scene match rate (i.e., H/C) for the NDNAD was 65.5% over the 2018–2020 period (National DNA Database Strategy Board, 2020). This value represents the fraction of crime scene profiles loaded onto the NDNAD that matched a subject/reference profile. The main crime types associated with these matches are burglary, vehicle crime, criminal damage, violent crime, drugs offences, robbery, murder and manslaughter, rape, theft and traffic offences (National DNA Database Strategy Board, 2020). It must be remembered, of course, that the only insight from the H/C rate is the potential of the NDNAD in solving 'DNA-related' crimes, thus contributing little to the policing of the great majority of crimes that occur, due to under-reporting and recording practices. For example, estimates by the Crime Survey for England and Wales (CSEW) show that whilst around 12 million crimes were committed in 2020, only about 5.6 million were recorded by the police (ONS, 2021).

The H/N calculates the fraction of subject (person) profiles loaded in a specific year that matched a crime stain profile already stored in the database. Figure 11.1 presents the H/N from April 2003 to March 2017, showing that at its peak (the 'DNA Expansion Programme' years, when large sums were spent on DNA profiling), less than 3.5% of subject profiles loaded matched with a crime scene profile, with the number normally hovering under 2%.

This H/N data suggests that both the ethical/privacy and financial costs of the NDNAD (around £2.5 million a year) must include an account of the annual sampling of the 98% of individuals who may not be linked to existing unsolved crimes or, indeed, any future crimes.

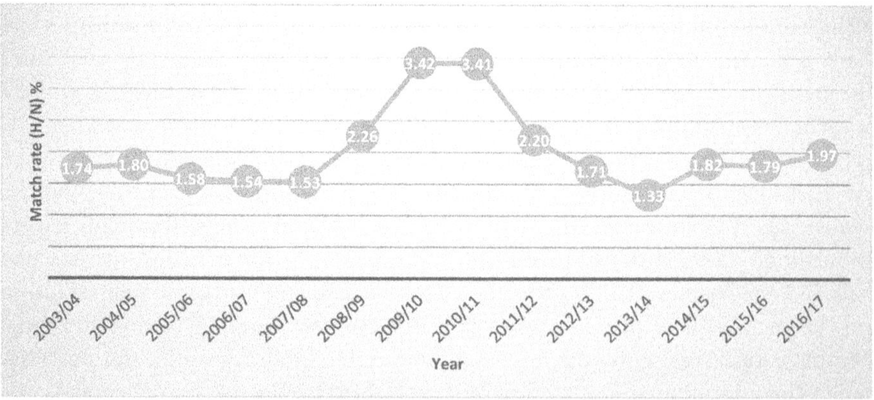

FIGURE 11.1 Trend of annual subject load match rate H/N (% loaded reference to crime stain matches per the number of annual subject/reference profiles loaded) 2003–2017

Source: Forensic Information Databases Service (formerly NDNAD Delivery Unit)

11.3.3 Perceived effectiveness

Perceived effectiveness examines the attitudes of members of the public regarding the impact of DNA databases – an indicator of the legitimacy of a database and its acceptability in society, which are critical to maintaining the integrity of a DNA database. The UK Home Secretary, at the time of the Marper ruling, flagged the importance of the public view, explicitly stating, 'Storing indefinitely the DNA and fingerprints of more than 1 million innocent people undermines public trust in policing and goes against any sense of natural justice' (Hansard, HC Deb., vol. 524, col. 205, 1 March 2011). The determination of appropriate law or policies to govern DNA databases has been described as a societal choice (Patyn and Dierickx, 2010). This choice (including judicial decisions) is influenced by claims of public security benefits, emphasising the need for realistic evidence of their effectiveness so that these 'societal choices' can be informed. As the Independent Advisory Group on the Use of Biometric Data in Scotland asserted:

> There is a need for an effective assessment of the benefit of these technologies to ensure that any new regime is based on utility and public safety and derives from sound evidence rather than anecdote or impression. It is also crucial to ensure that there is greater transparency and public participation around the use of biometric data in the criminal context.
>
> *(IAG on the Use of Biometric Data in Scotland, 2018:40)*

Although there have been attempts in several countries to assess the public acceptance of DNA databases (e.g., Carvalho (2007); see Amankwaa (2018) and Machado and Silva (2019) for reviews), there is presently no assessment of public views on the DNA database in the UK and no mention of such indicators by oversight bodies. In a public survey, Amankwaa (2019) found respondents were equivocal about the ability of the NDNAD to prevent crime but believed the NDNAD impacts crime detection, prosecution and investigation, with over half viewing these benefits as outweighing civil liberty concerns. When a sample of law enforcement officers was surveyed, the majority believed expanding the NDNAD would make it more effective (Amankwaa, 2019), mirroring the high support for DNA databases among law enforcement found by Teodorović *et al.* (2017) in Serbia.

Further research would be required to demonstrate whether the UK NDNAD is meeting the expectations of the public, having their confidence and trust. To account for the social acceptability of DNA databases, a biometric perception index (BPI) backed by oversight bodies and statute could be instigated (Amankwaa, 2019; Amankwaa and McCartney, 2019). The BPI would periodically survey the perceived effectiveness of DNA databases and views on inclusion and retention criteria among a representative sample of the population and be reported by oversight bodies in their annual reports. This information would be invaluable in supporting arguments concerning the legitimacy and acceptability of DNA databases, two important criteria that should be incorporated into expanded evaluations of DNA database effectiveness.

11.4 Expanding effectiveness: The 'integrity' of DNA databases

Current evaluative criteria then do not achieve a realistic and holistic view of DNA databasing, for 'the pursuit of justice means more than simply the resolution and reduction of crime' (Krimsky and Simoncelli, 2011: xvii). To ensure an ethical criminal justice process, respectful of human rights and based on socially accepted notions of 'justice',[2] DNA database operation and governance must have 'integrity'. The leading case (at least in Europe) of *S and Marper* (2008) demanded a change of approach away from 'effectiveness' to incorporate a human rights focus when states seek to justify their powers to take and retain the DNA of citizens. Yet, as was clear when the UK government were considering their legislative response to Marper, there was 'no attempt at all to develop a human rights-based approach to retention' (McCartney, 2012:251; see Skinner and Wienroth, 2019). This has been contrasted with the Netherlands, where respect for individual rights and legal principles was the starting point for creation of their DNA database (see Toom, 2012).

Often, debate has centred on the make-up of the database, focusing on the over-representation of ethnic minorities, as well as other demographic characteristics, such as gender and age. There have been suggestions that the introduction of a universal database would neutralise the skewed composition of databases and their privacy implications for the over-represented (Williams and Johnson, 2004; Hazel *et al.*, 2018). However, the NDNAD is an output of policing practices, statutory requirements, justice policy and the nature or context of crime (Skinner, 2013; Skinner and Wienroth, 2019). Further, it is unlikely, given judicial decisions specifically in relation to state retention of DNA in databases, that a universal database would ever be found 'necessary' or 'proportionate'.[3]

Concerns over the skewed nature of databases, albeit important, should not be the singular focus of debate. In other related domains, it is accepted that issues surrounding new technologies and the state use of information, for example, include social, ethical and legal dimensions. In 2007, the Nuffield Council on Bioethics published a critical report on the forensic use of bioinformation, focusing on those principles to be respected, such as the respect for personal liberty, the maintenance of the autonomy of the individual, personal privacy, informed consent and equal treatment (Nuffield Council on Bioethics, 2007). More recently, Scotland has created a legislative framework for the police use of biometrics, with the Scottish Biometrics Commissioner Act 2020, section 2 (1) stating that the function of the new commissioner is to support and promote the adoption of 'lawful, effective and ethical practices in relation to the acquisition, retention, use and destruction of biometric data for criminal justice and police purposes'.[4] While 'lawful' might invoke the traditional legal tests of necessity and proportionality, it still raises the question of how the commissioner is going to 'support and promote *effective* and *ethical*' practices (emphasis added). Using which measurements and against what criteria will 'effective' and 'ethical' be gauged? Meanwhile, the newly appointed UK biometrics commissioner[5] stated in an interview upon appointment:

> I think this area is all about balancing three things: what we can do technically, what can be done legally, and more broadly what will get the support of communities and society at large. Getting that balance right is the strategic challenge ahead.
>
> *(Biometrics Commissioner, 2021)*

There have been warnings about the lack of a proper legal basis for many biometric technologies and collections across Europe (e.g., Kindt, 2018) and suggestions for practical frameworks encompassing ethical and social measures to be applied to biometric technologies. In an echo of the exhortation by the UK government's chief scientific advisor (2015:10) that 'We can only have the best discussion about innovations if we understand that the discussion must be about both science and values', Wienroth (2020:92) stresses a need to demonstrate 'value beyond validity', proposing 'a practice-based approach to testing values in new technologies and their respective emerging practice and governance arrangements around Reliability, Utility and LEgitimacy (RULE)'. McCartney (2015, 2017b) argues that DNA databases, their operation and their governance should have 'integrity' as their overarching evaluative criterion. Integrity is essential for trust among not just the 'users' of forensic DNA data generated but also the wider public (particularly if they have a role in the criminal justice process). The question then becomes, 'How do we ensure integrity in forensic DNA databases?' There are clear commonalities with other regulatory models, 'codes of practice' and guidelines that proliferate around not just DNA but other biometrics and technologies such as AI. Criteria for (anticipatory) governance (e.g., Oswald (2021), Wienroth (2020)), as well as terms of reference for oversight bodies, include many of the same or similar benchmarks. Thus, a DNA database should be assessed against a matrix of standards to be met or achieved (see Figure 11.2), which, when combined, would provide a holistic evaluation of the 'integrity' of a DNA database.

Viability should be assured by guarantees that the database 'works' as it should (technical operation) and that data gathered and produced is reliable and can be widely understood without significant 'translation' or complex interpretation required, which could lead to confusion and variability in results (universality). The operation of the entire system and processes should be guaranteed (as far as possible) by quality assurance mechanisms that check the system against agreed standards. The notion of a database that 'works' is that focused on with current measures around 'effectiveness', although, as previously discussed, this is only a small sub-set of 'viability' as outlined here.

The second pillar is that of legitimacy: first and foremost, the database must be lawful, and no operations can take place (without detection and sanctions) outside the laws that govern the database. These laws must ensure that the database is both

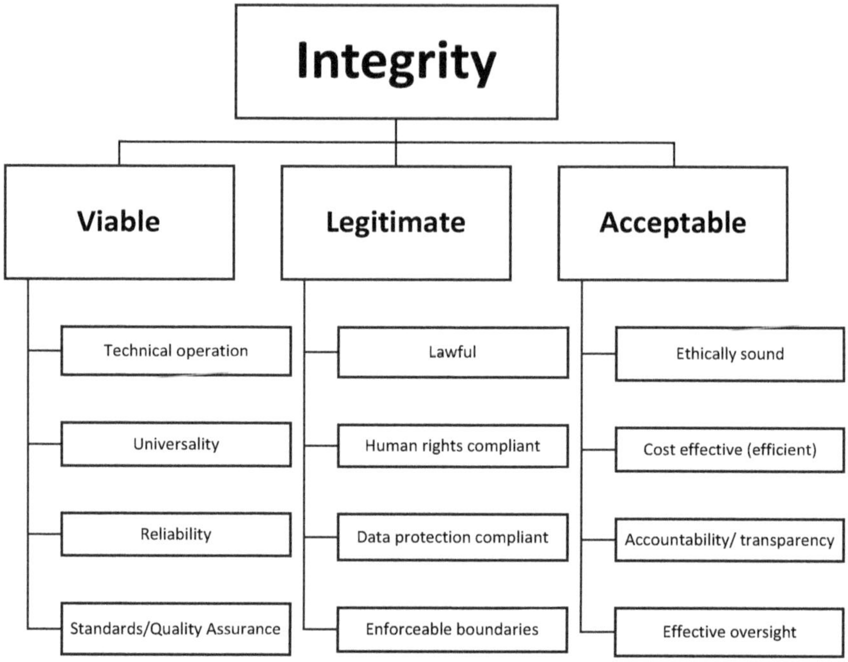

FIGURE 11.2 The integrity matrix

human rights and data protection compliant. This should ensure that privacy rights are properly accounted for and that the database can be justified as 'necessary and proportionate' under human rights laws. These laws and the operation of the database must exist within strict 'boundaries' that are enforceable, meaning that the data cannot be shared without lawful permission, and the database cannot be used for non-permitted purposes, preventing abuse of the data/database and requiring further debate and legal changes before 'mission creep' can occur.

Thirdly, databases must be socially acceptable (the 'perceived effectiveness' discussed earlier); most obviously, this must encompass respecting relevant ethical principles. There must also be sufficient information for the public and policy makers to make an assessment of the (cost) effectiveness of the DNA database, using a 'dissected' approach to evaluation outlined earlier, with reliance on the potential effectiveness parameters, H/C and H/N, to assess the capacity of DNA databases to contribute to crime solving and the representativeness of reference databases, respectively. A robust research programme should be developed to assess the end-to-end probative value of NDNAD hits. This latter assessment should focus on at least the initial role of databases in generating probative cold hits. These hits must then be tracked through the criminal process to determine their impact on public security outcomes and indicators. Only after such cumulative evaluations can inclusion and retention criteria be identified to create the most effective composition of the database. There can then be proper consideration of the costs and benefits of running DNA databases in order to make evidence-based decisions on their parameters: 'Once we reach an accurate assessment what has been achieved – we can plot a path to exploit potential' (Doleac, 2017). Statutory oversight bodies must have adequate capacity and comprehensive powers to ensure effective oversight and governance. The implementation of these recommendations may improve the transparency and public confidence in the governance and operation of DNA databases, crucial to their 'acceptability'. 'Integrity' is then secured by ensuring that a DNA database is viable, legitimate and acceptable.

11.5 Conclusion: A holistic approach to evaluating DNA databases?

The advent of forensic DNA may have been revolutionary, leading to the accurate detection and conviction of many criminals who might otherwise have evaded punishment. Although a probative DNA database match in a single case can be priceless (van der Beek, 2015), the impact of mass-retention of DNA of citizens in large databases requires further interrogation:

> [T]he benefits of mass DNA retention can be shown in a very case specific manner. In terms of individual rights it is very difficult to argue against the limited impact upon a person that the taking of their DNA will have when pitted against such a benefit. What is more difficult to gauge however, is the wider cumulative societal impact that DNA retention will have.
>
> *(Crossman et al., 2007:10)*

The main drivers for the expansion of DNA databases internationally may be their high perceived effectiveness among members of the public, but their support may have been secured with potential misinterpretation or misunderstanding of (scarce and elementary) available data and individual success stories in high-profile cases. Yet while DNA databases may offer slightly improved detection or conviction rates in some specific crime types, their overall contribution to public security outcomes is severely limited. Most crimes are never reported, and the overwhelming majority of recorded crimes do not involve DNA. The utility of any DNA match is also entirely dependent on the competency and efficiency of the police, prosecutors and other participants in the criminal justice process. This all means that DNA databases should be considered a highly specialised investigative tool while the privacy, ethical and financial costs of operating DNA databases requires that they are developed progressively. Decisions regarding their constitution must be based on realistic evidence, rather than pursuing an expansionist agenda supported by over-inflated accounts of benefits that could accrue. As explained by the IAG on the Use of Biometric Data in Scotland (2018):

> Ultimately, the public will have a number of choices to make about the type of society in which they wish to live. There is always a balance to be struck between, on the one hand, considerations of public protection and, on the

> other, the right to privacy and other relevant human rights and ethical considerations. It can be difficult to have a rational debate in the aftermath of specific news stories which may emphasise only one part of the argument. There needs to be a wider debate about the various implications of the capture or surrender of biometric data, especially in terms of the implications for privacy.
> *(p. 11)*

While DNA has been referred to using many hyperbolic epithets, scrutiny soon reveals that, as with any technology, it is not infallible and may even lead to injustice. It has previously been rare that an ethical spotlight has been shone on forensic science; indeed, it has customarily been considered unproblematically in the 'public good', and powers to take and retain personal biometrics in large collections date back to the birth of fingerprinting. States and police agencies have habitually been meticulous record keepers, and collecting and using personal biometric data were originally seen as mere extensions of such accepted powers. However, for a variety of reasons, the 'innocent have nothing to fear' argument has been losing potency. There is now a greater realisation that there may be social and ethical consequences to forensic technologies and the police powers required to utilise them:

> As in so many areas, it is dangerous to pursue purely technological solutions to human challenges. By itself, technology cannot solve anything. Everything depends on its appropriate and acceptable use and that requires implicit trust between the authorities and the people. If the very use of technology erodes that trust it will only worsen the problems it is designed to solve.
> *(Thornhill, 2019)*

In absence of data and broader account of the 'integrity' of a database, optimal scale and arrangements cannot be known, leaving undetermined where any 'balance' may lie and asking the public for perhaps underserved trust or 'blind faith'. Genetic privacy is popularly portrayed as being 'traded' for public security outcomes, but this is not informed by evidence. While there are, of course, advantages to building DNA databases that are deemed 'worth' sacrificing some civil liberties for, presently (and maybe always?), public security outcomes appear limited, and most remain highly speculative. Further, the 'balance' metaphor offers no resources for determining what the 'right' balance looks like and assumes (wrongly?) that any extension of individual privacy rights compromises 'public safety' on the other side of the scales. It is becoming clear that this is not how security/privacy equations work, with Solove (2011) and others powerfully arguing that increasing privacy for individuals actually creates a more secure society. Any 'balance' struck will also always be transitory (subject to alteration when uses change), contentious and culturally specific, so it cannot simultaneously also be prescriptive and universally accepted.

Questions of whether states are collating excessive information on citizens will persist, particularly where information could be used for discriminatory purposes and to the detriment of groups or individuals. At a national level, (bio)surveillance of citizens could lead to a 'suspect society' (McCartney, 2006, 2007; Cole and Lynch, 2006; Lynch and McNally, 2013). This may occur by stealth, with police powers subtly extended, new opportunities arising for DNA sampling and new technological developments in forensic genetics. A coherent and honest assessment and broader evaluative strategy are also vital when other biometric technologies attempt to 'piggyback' on the apparent 'success' of DNA databases. A realistic and holistic weighing of the benefits brought by forensic DNA databasing necessarily involves reflection on mistakes of the past and consideration of whether there are now the foresight, ability and will to prevent abuses and over-reach and augment the advantages of forensic DNA databases in a viable, legitimate and socially acceptable fashion.

Notes

1 For just two well-known examples, you can look to the famous cases of the 'Phantom of Heilbronn' in Germany and the investigation into the 'body in the bag' death of MI6 employee Gareth Williams, but presumably, more occur than ever become public knowledge.
2 Clearly there are variances across individuals and society of exactly what constitutes 'justice', but generally, a democratic society will have a notional 'acceptance' of what the pursuit of justice permits with respect to state powers, etc.
3 Both the United Arab Emirates and Kuwait have attempted to launch 'universal' DNA databases and have failed, with strong criticism from human rights organisations and the UN Human Rights Committee (see Joly et al., 2019).
4 The Scottish biometrics commissioner thus has a different remit from that of the UK biometrics commissioner, whose statutory basis is found in the Protection of Freedoms Act 2012.
5 This new commissioner has also taken on the role of surveillance commissioner (these were previously separate roles), with the Home Office suggesting in 2021 that the role should also absorb the information commissioner remit. The new commissioner must tackle the challenges posed by the introduction of automated facial recognition, another area of policy that has been described as 'a hot mess' by the CEO of the Royal Statistical Society (Ward-Foxton, 2019).

References

Amankwaa, A. O. (2018) 'Forensic DNA retention: Public perspective studies in the United Kingdom and around the world', *Science & Justice*, 58(6), pp. 455–464. doi: 10.1016/j.scijus.2018.05.002.

Amankwaa, A. O. (2019) *Forensic DNA databasing: Retention regimes and efficacy*. PhD. University of Northumbria at Newcastle. Available at: http://nrl.northumbria.ac.uk/42060/1/Amankwaa.Aaron_phd.pdf.

Amankwaa, A. O. and McCartney, C. (2019) 'The effectiveness of the UK national DNA database', *Forensic Science International: Synergy*, 1, pp. 45–55. doi: 10.1016/j.fsisyn.2019.03.004.

Amankwaa, A. O. and McCartney, C. (2021) 'The effectiveness of the current use of forensic DNA in criminal investigations in England & Wales', *WIREs Forensic Science*, p. e1414.

Anker, A. S. T., Doleac, J. L. and Landersø, R. (2021). 'The effects of DNA databases on the deterrence and detection of offenders', *American Economic Journal: Applied Economics*, 13(4), pp. 194–225. doi: 10.1257/app.20190207.

Aronson, J. (2007) *Genetic witness: Science, law, and controversy in the marketing of DNA profiling*. New Brunswick, NJ: Rutgers University Press.

Bieber, F. R. (2006) 'Turning base hits into earned runs: Improving the effectiveness of forensic DNA data bank programs', *Journal of Law, Medicine & Ethics*, 34(2), pp. 222–233. doi: 10.1111/j.1748-720X.2006.00029.x.

Biometrics Commissioner (2021) 'Police should not be banned from using facial recognition technology, says UK watchdog', *The Financial Times*. Available at: www.ft.com/content/79223f6e-a772-4e74-b256-88641a416f92?shareType=nongift (Accessed: 20 May 2021).

Black, S. (2019) *All that remains*. Ealing, London: Transworld Publishers.

Bramley, B. (2009) 'DNA databases', in Fraser, J. and Williams, R. (eds) *Handbook of forensic science*. Cullompton: Willan Publishing, pp. 309–336.

Business Wire India (2021) 'Offender DNA database will be strong deterrent to repeat crimes in India', 13 April. Available at: www.businesswireindia.com/offender-dna-database-will-be-strong-deterrent-to-repeat-crimes-in-india-72499.html (Accessed: 28 April 2021).

Carvalho, M. (2007) 'Spanish public awareness regarding DNA profile databases in forensic genetics: What type of DNA profiles should be included', *Journal of Medical Ethics*, 33(10), pp. 598–604.

Cole, S. A. and Lynch, M. (2006) 'The social and legal construction of suspects', *Annual Review of Law and Social Science*, 2(1), pp. 39–60.

Crossman, G., Kitchin, H., Kuna, R., Skrein, M. and Russell, J. (2007) *Overlooked: Surveillance and personal privacy in modern Britain*. London: Liberty.

Doleac, J. L. (2017) 'The effects of DNA databases on crime', *American Economic Journal: Applied Economics*, 9(1), pp. 165–201. https://doi.org/10.1257/app.20150043.

European Commission (2011) *Crime and deviance in the EU—key findings from EU funded social sciences and humanities research projects*. Luxembourg: Publications Office of the EU.

FIND Strategy Board (2020) *National DNA database strategy board biennial report 2018–2020*. London: Forensic Information Database Strategy Board.

Fraser, J. (2020) *Murder under the microscope*. London: Atlantic Books.

Goulka, J. *et al.* (2010) *Toward a Comparison of DNA Profiling and Databases in the United States and England*. Santa Monica, CA: RAND Corporation, p. 34.

Hazel, J. W. *et al.* (2018) 'Is it time for a universal genetic forensic database?' *Science*, 362(6417), pp. 898–900. doi: 10.1126/science.aav5475.

HMIC (2012) *Policing in austerity: One year on*. London: Her Majesty's Inspectorate of Constabulary. Available at: www.justiceinspectorates.gov.uk/hmicfrs/media/policing-in-austerity-one-year-on.pdf (Accessed: 30 January 2020).

Home Office (2009) *Keeping the right people on the DNA database: Science and public protection*. London: Home Office.

Home Office (2021) 'Crime outcomes in England and Wales, year to December 2020: Data tables'. Available at: www.gov.uk/government/statistics/crime-outcomes-in-england-and-wales-year-to-december-2020-data-tables (Accessed: 16 May 2021).

Hough, M. and Sato, M. (2011) *Trust in justice: Why it is important for criminal policy, and how it can be measured*, Final report of the Euro-Justis project. Helsinki: HEUNI.

Human Genetics Commission (2009) *Response to the government DNA consultation*. London: Human Genetics Commission.

Hunter, J. P. R., Young, S. and Dixon, D. (eds) (2016) *The integrity of criminal process: From theory into practice*. Oxford: Hart Publishing.

IAG (2018) *Report of the independent advisory group on the use of biometric data in Scotland*. Edinburgh: Scottish Government. Available at: www.ohchr.org/Documents/Issues/RuleOfLaw/PeacefulProtest/NHRIs/scottish-human-rights-commission4.pdf (Accessed: 20 May 2021).

Joly, Y., Marrocco, G. and Dupras, C. (2019) 'Risks of compulsory genetic databases', *Science*, 363(6430), pp. 938–940. doi: 10.1126/science.aaw4347.

Kindt, (2018) '"Having yes, using no." About the new legal regime for biometric data', *Computer Law and Security Review*, 34(3), pp. 523–538.

Krimsky, S. and Simoncelli, T. (2011) *Genetic justice: DNA data banks, criminal investigations, and civil liberties*. New York: Columbia University Press.

Leary, D. and Pease, K. (2003) 'DNA and the active criminal population', *Crime Prevention & Community Safety*, 5(1), pp. 7–12. doi: 10.1057/palgrave.cpcs.8140135.

Lynch, M. and McNally, R. (2013) 'Forensic DNA databases and biolegality: The co-production of law, surveillance technology and suspect bodies', in Atkinson, G. and Lock, M. (eds) *Handbook of genetics and society: Mapping the new genomic era*. London: Routledge.

Machado, H. and Silva, S. (2019) 'What influences public views on forensic DNA testing in the criminal field? A scoping review of quantitative evidence', *Human Genomics*, 13, Article 23.

McCartney, C. (2006) 'The DNA expansion programme and criminal investigation', *British Journal of Criminology*, 46(2), pp. 175–192. doi: 10.1093/bjc/azi094.

McCartney, C. (2007) *Forensic identification and criminal justice: Forensic science, justice and risk*. Cullompton: Willan Publishing.

McCartney, C. (2012) 'Of weighty reasons and indiscriminate blankets: The retention of DNA for forensic purposes', *The Howard Journal*, 51(3), pp. 245–260.

McCartney, C. (2015) 'Forensic data exchange: Ensuring integrity', *Australian Journal of Forensic Sciences*, 47(1), pp. 36–48.

McCartney, C. (2017a) 'DNA and identification', in Maguire, M. and Holt, T. J. (eds) *Handbook of technology, crime and justice*. London: Routledge.

McCartney, C. (2017b) 'Trust and the international exchange of forensic information', in Hufnagel, S. and McCartney, C. (eds) *Trust in international police and justice cooperation*. Oxford: Hart Publishing.

National DNA Database Strategy Board (2020) *National DNA database: Biennial report, 2018 to 2020*. London: HMSO.

North Yorkshire Police (2018) 'Cold case DNA breakthrough secures conviction of 1988 York rapist', *North Yorkshire Police*, 21 March. Available at: https://northyorkshire.police.uk/news/cold-case-dna-breakthrough-secures-conviction-1988-york-rapist/ (Accessed: 13 December 2019).

Nuffield Council on Bioethics (2007) *The Forensic Use of Bioinformation: Ethical Issues*. Nuffield Council on Bioethics, London. Available online at: https://www.nuffieldbioethics.org/wp-content/uploads/The-forensic-use-of-bioinformation-ethical-issues.pdf.

Office for National Statistics (ONS) (2021) *Crime in England and Wales: Year ending December 2020*. Available at: www.ons.gov.uk/peoplepopulationandcommunity/crimeandjustice/bulletins/crimeinenglandandwales/yearendingdecember2020.

Oswald, M. (2021) 'A three-pillar approach to achieving trustworthy and accountable use of AI and emerging technology in policing in England and Wales: Lessons from the west midlands model'. http://dx.doi.org/10.2139/ssrn.3812576.

Patyn, A. and Dierickx, K. (2010) 'Forensic DNA databases: Genetic testing as a societal choice', *Journal of Medical Ethics; London*, 36(5), p. 319. doi: http://dx.doi.org/10.1136/jme.2009.033829.

Pearson, E. (2018) 'Victorians under arrest to be DNA swabbed under new legislation', *The Age*. Available at: www.theage.com.au/national/victoria/victorians-under-arrest-to-be-dna-swabbed-under-new-legislation-20180619-p4zmet.html (Accessed: 8 August 2018).

Prainsack, B. and Toom, V. (2013) 'Performing the union: The Prüm decision and the European dream', *Studies in History and Philosophy of Biological and Biomedical Sciences*, 44, pp. 71–79.

Skinner, D. (2013) '"The NDNAD has no ability in itself to be discriminatory": Ethnicity and the governance of the UK national DNA database', *Sociology*, 47(5), pp. 976–992. doi: 10.1177/0038038513493539.

Skinner, D. and Wienroth, M. (2019) 'Was this an ending? The destruction of samples and deletion of records from the UK police national DNA database', *BJHS Themes*, pp. 1–23. doi: 10.1017/bjt.2019.7.

Solove, D. (2011) *Nothing to hide: The false tradeoff between privacy and security*. New Haven, CT: Yale University Press.

Struyf, P. *et al.* (2019) 'The effectiveness of DNA databases in relation to their purpose and content: A systematic review', *Forensic Science International*, 301, pp. 371–381. doi: 10.1016/j.forsciint.2019.05.052.

TASS (Russian News Agency) (2021) 'DNA database helped to solve 30,000 crimes in Russia since 2009—interior ministry', 8 May. Available at: https://tass.com/russia/1287529 (Accessed: 9 May 2021).

Teodorović, S., Mijović, D., Radovanović Nenadić, U. and Savić, M. (2017) 'Attitudes regarding the national forensic DNA database: Survey data from the general public, prison inmates and prosecutors' offices in the Republic of Serbia', *Forensic Science International: Genetics*, 28, pp. 44–51. doi: 10.1016/j.fsigen.2017.01.007.

Thornhill, J. (2019) 'Regulate facial recognition technology before it is too late', *The Financial Times*, 27 June.

Toom, V. (2012) 'Forensic DNA databases in England and the Netherlands: Governance, structure and performance compared', *New Genetics and Society*, 31(3), pp. 311–322.

Toom, V., Granja, R. and Ludwig, A. (2019) 'The Prüm decisions as an aspirational regime: Reviewing a decade of cross-border exchange and comparison of forensic DNA data', *Forensic Science International: Genetics*, 41, pp. 50–57.

UK Govt Chief Scientific Advisor, Annual Report (2015) *Forensic science and beyond*. London: HMSO.

van der Beek, K. (2015) 'Measuring the effectiveness and efficiency of forensic DNA-databases', in *International symposium on human identification (ISHI)*. Grapevine, TX: Promega Corporation. Available at: www.promega.com/-/media/files/products-and-services/genetic-identity/ishi-26-oral-abstracts/9-van-der-beek.pdf (Accessed: 9 April 2017).

Walsh, S. J., Curran, J. M. and Buckleton, J. S. (2010) 'Modeling forensic DNA database performance', *Journal of Forensic Sciences*, 55(5), pp. 1174–1183. doi: 10.1111/j.1556-4029.2010.01426.x.

Ward-Foxton, S. (2019) 'Facial recognition: The ugly truth', *EE Times Europe*, 30 September. Available at: www.eetimes.eu/facial-recognition-the-ugly-truth/.

Wienroth, M (2020) 'Value beyond scientific validity: Let's RULE (Reliability, Utility, LEgitimacy)', *Journal of Responsible Innovation*, 7(Sup 1), pp. 92–103. doi: 10.1080/23299460.2020.183515.

Wiles, P. (2017) *Annual report 2016: Commissioner for the retention and use of biometric material*. London: Office of the Biometrics Commissioner.

Wiles, P. (2019) *Annual report 2018: Commissioner for the retention and use of biometric material*. London: Office of the Biometrics Commissioner.

Williams, R. and Johnson, P. (2004) '"Wonderment and dread": Representations of DNA in ethical disputes about forensic DNA databases', *New Genetics and Society*, 23(2), pp. 205–223. doi: 10.1080/1463677042000237035.

12

THE STAKES OF FORENSIC PHENOTYPIC PROFILING

Can solidarity help?

Barbara Prainsack and Gabrielle Samuel

Introduction

Since the last decade of the 20th century, the dominant method of forensic DNA analysis – so-called short tandem repeat (STR) profiling – has been to compare a specific set of DNA markers from an unidentified crime scene sample with the markers from a DNA sample of a suspected perpetrator to determine whether they 'match'. Two profiles match if the tested markers in the same locations look the same – meaning that the suspected perpetrator can be linked to the crime scene. However, there may be very legitimate reasons for the suspected perpetrator to have left their DNA; the genetic match says nothing about the kind of connection that the person has to the crime. Further evidence would be required to determine whether the suspect is, in fact, the perpetrator, such as witness statements, confessions and other marks left at the crime scene.

The use of DNA profiling in the criminal justice system has raised several questions about how the benefits of the technology – ideally, the successful identification of a perpetrator – can be balanced with potential intrusions of civil rights, such as privacy, discrimination and surveillance (e.g., Hindmarsh and Prainsack 2010; Chow-White and Duster 2011; Toom 2012; Wienroth et al. 2014; Wallace et al. 2014; Ahuriri-Driscoll et al. 2021). For a long time, the distinction between coding and non-coding DNA played a useful role as an ethical boundary. It was understood that the analysis of non-coding regions can – via association with coding regions – also provide information on internal or external traits of people (Samuel and Prainsack 2018b). By deliberately avoiding the analysis of sections of DNA within the coding region, and by only analysing specific non-coding STR sites, this problem was largely bypassed. In effect, the coding versus non-coding distinction had obtained technolegal normativity by separating (supposedly) ethically sensitive applications of DNA-based technologies from those that were seen as

DOI: 10.4324/9780429322358-16

less ethically problematic. Recently, greater use of newer methods of genetic analysis, fuelled in part by technological advances such as massive parallel sequencing (see Chapter 5, this volume), has troubled this ethical boundary. As we will show in this chapter, some (though not all)[1] of the newer technological practices deliberately look at both non-coding and coding regions to infer, through probabilistic estimates, what a perpetrator may look like.

Forensic DNA phenotyping: New technolegal normativities

Forensic DNA phenotyping (FDP), also sometimes called 'extended DNA analysis' (see Amelung et al. 2021), is one such technological practice that specifically looks at coding regions of the DNA. It does so to help identify a perpetrator in cases in which clues to the identity of the perpetrator or their appearance are unavailable. The purpose of FDP is thus different from traditional STR-based forensic DNA analysis. FDP, by its very definition, aims to analyse a sample of an unknown person's DNA to say something useful about that person's observable, external characteristics – which STR profiling seeks to avoid (see also Granja and Machado 2020). Broadly speaking, and for the context of this chapter, we can consider the phenotypic tests for age, biogeographical ancestry and appearance as falling within the remit of FDP. These include appearance traits – so-called externally visible characteristics (EVC), such as eye, hair or skin colour – and markers for biogeographical ancestry.

FDP for the prediction of appearance traits or biogeographical ancestry requires the analysis of hundreds of single nucleotide polymorphisms (SNP), which are sites in a DNA sequence where a single base[2] varies between individuals. FDP can also require the analysis of other polymorphisms such as indels, which are naturally occurring additions or deletions in people's DNA. For the prediction of age analyses, FDP looks at DNA methylation[3] patterns at specific sites on the genome that have been found to change with age. Both types of FDP findings are probabilistic, meaning that they can infer a specific phenotypic feature only to a certain degree of probability. Once a suspect is identified with the help of FDP, a 'traditional' STR-based profile derived from this suspect is compared with a crime scene profile to see whether the two profiles match.

There is great variation across countries as to whether FDP is considered lawful in a given jurisdiction and whether it is practiced (Samuel and Prainsack 2018b). The legal assessment is complicated by the fact that most countries do not have legislation that explicitly addresses FDP. This is because they issued legislation on the use of DNA material and data for criminal investigation and forensics when STR profiling first became widespread, and DNA databases were set up in the late 1990s and early 2000s. At that time, the ethical relevance of the coding versus non-coding distinction was firmly intact in public discourse and among experts. At that time, legislators did not foresee the development of technologies that would use DNA for purposes other than identification via STR matches – and that they would thus need to look also at coding DNA. As noted, the coding versus non-coding distinction had strong technolegal normativity at this time, and strong regulatory practices were established around it. In some jurisdictions, the legitimate use of DNA material for analysis by law enforcement was limited to non-coding regions of the genome (Samuel and Prainsack 2018b). Because FDP markers are, by definition, located in both non-coding and coding regions, with no distinction between the two, this effectively (yet implicitly) outlawed FDP.

In Europe, few countries currently have explicit legislation pertaining to FDP use in the criminal justice system. Countries are slowly moving to formulate new legislation to try and address the issue. This means that the regulatory picture is continually changing. For the countries that do permit the use of FDP, it is not always for all phenotypic traits.[4] These different legal situations exemplify the different understandings of policy makers regarding the ethical and societal sensitivities associated with each of the phenotypic traits (see also Schneider et al. 2019).

Supporters of FDP purport that this technology is uniquely placed to provide leads in criminal cases where no other leads are available. If it is paired with suitable safety and accountability measures, they argue, societies have an ethical obligation to honour the commitment to try to solve crimes and to enhance the security of their members (e.g., Kayser 2015; for a discussion of these arguments, see M'charek et al. 2012; Wienroth 2018). For others, the challenges that FDP poses outweigh the benefits that could be achieved by its use. An emerging social science literature has discussed these challenges, and we review them in detail elsewhere (Samuel and Prainsack 2019b; see also Granja and Machado 2020). The key concerns articulated in this literature pertain to the risk of discriminating against minority groups, infringements on privacy and autonomy and the misinterpretation of FDP findings by various stakeholders. In the following sections, we will summarise these key concerns. We will also explore new ethical approaches to help us regulate and use FDP in a responsible manner and reflect on what this means in terms of technolegal practices. Specifically, FDP challenges us to develop new 'ethical boundaries' to help balance its utility for law enforcement (and, ideally, for security as a public good) with other societal and ethical concerns, including excessive surveillance and infringements on civil liberties. 'Balancing' here is not to be understood as a process by which one goal or good has to be compromised in order for another one to gain weight; we fully support the view that the relationship between security and privacy is not a zero-sum game in which one can be 'traded' for another (Pavone et al. 2016). Instead, we need to explore whether technological practices that are employed with the explicit goal of enhancing security do, in fact, address security needs and how they could ideally enhance other goods such as privacy and autonomy (Samuel and Prainsack 2019a). Technolegal normativities of FDP are already forming, most prominently focusing on the ethical acceptability of specific phenotypic characteristics rather than – as is the case with STR testing – the acceptability of the specific type of DNA segment (coding or non-coding) being analysed. We explore the underlying assumptions associated with the ethical boundaries pertaining to these phenotypic characteristics. In doing so, finally, we will argue that solidarity can help us develop an approach that acknowledges the potential benefits of FDP while taking seriously – and seeking to minimise – the risks that it poses, especially for minorities in our societies.

Correct analysis and interpretation of FDP findings

There are concerns about the correct interpretation of FDP information. Phenotypic traits are multifactorial, meaning that they are not determined by one gene alone but by a complex interplay between a multitude of genetic and environmental factors. This means that genetic data from FDP tests can only be used to 'predict' a person's phenotypic characteristics to a certain degree of likelihood. Because of this, several scholars have raised concerns about the nature of information FDP can provide and the possibility of such information being overinflated. They are especially worried that practitioners who do not normally deal with probabilistic information and thus may not be accustomed to the type of information yielded by FDP testing may misinterpret FDP findings (Enserink 2011; Cino 2017; Sankar 2012; Seo et al. 2017; Toom et al. 2016). Moreover, many scholars have noted that the accuracy of a prediction based on crime scene DNA is also contingent on many variables related to the collection and preservation of DNA, the likelihood of contamination, the type of genetic markers used during FDP (Cino 2017; Sankar 2012; Seo et al. 2017) and the risks of false positive and false negative test results (Haga 2006).

Discrimination against minorities and marginalised populations

A range of scholars and experts from various fields of science and practice are concerned that FDP could be used in ways that discriminate against, or stigmatise, persons and groups (e.g., Koops et al. 2006). This is particularly the case for biogeographical ancestry testing. For instance, if FDP were used in such a way that emphasised dividing lines around ethnic, religious or 'cultural' groups, such that certain groups of people were portrayed as different from the rest of society (e.g., people with genetic ancestry from certain world regions), then this would have a negative effect on social cohesion and could reinforce existing divisions and prejudices in our society. The risk of stigmatising groups would be particularly high if these groups were associated with a higher prevalence of crime in public discourse (Grewal 2007; Hagan et al. 2008; M'charek 2008). It could also be exacerbated by structural racism in our societies, where assumed differences and hierarchies between groups of people defined along ethnic or 'racial' lines are inscribed in our societal and political institutions and shared practices (Samuel and Prainsack 2019b; M'charek et al. 2020).

Against this backdrop, discrimination could take place in many places and forms. These include the misinterpretations of FDP findings by police officers, which some social science scholars argue could increase racial profiling. Here, FDP predictions could be interpreted as supporting deep-seated individual or structural biases against specific ethnic or other minority groups. If FDP findings were released to the public, they could also cause a divide between communities under suspicion and other groups within society. Finally, social science and ethics scholars have voiced concern that the use of FDP could lead to the reification of the belief in the genetic basis of race, which might, in turn, increase discrimination (e.g., Duster 2005; M'charek et al. 2014; M'charek 2020). While biogeographical ancestry is not seen as equivalent to race, these scholars argue that the technology's premise is still firmly rooted in creating genetic categories of difference using population genetic research, much like that of the concept of race (Sankar 2010; Dewey-Hagborg 2015).

Privacy and autonomy

Many scholars and experts raise concerns about privacy rights in the context of FDP. In the case of 'traditional' STR-based DNA analysis, threats to privacy rights are seen as less acute because STR profiles do not disclose any information about external or internal traits of people. Regarding FDP, scholars have discussed the possibility of privacy infringements in situations where the image that a person chooses to present to the world is not reflective of her genetic endowments (e.g., Ossorio 2006). For example, informing a person that their genetic ancestry is likely to be Northern European may conflict with the person's (and their family's) understanding of origin from another part of the world. In other cases, if a person has chosen to change their appearance – such as by wearing contact lenses, dying their hair or undergoing cosmetic treatments – some scholars have argued that being confronted with the DNA-based inferences of their likely appearance may conflict not only with the person's privacy but also with their autonomy, understood as the ability to be the author of their own life (Ossorio 2006). An additional concern stems from the fact that phenotypic testing for a trait that is not normally considered particularly sensitive – such as eye colour – can also disclose information about a much more sensitive trait (Weichert 2017; Haga 2006; Koops and Schellekens 2008). For example, it is known that people with fair skin and red hair are at higher risk of melanoma because of an increased sensitivity to sunlight (Ransohoff et al. 2016). We add that, when used in actual casework, FDP could interfere with people's privacy by including people in the investigation (as suspects, witnesses or family members or friends of suspects) who would otherwise not be included. Although such inclusions of innocent people in criminal investigations are routine occurrences in the criminal justice system that cannot be abolished entirely, if FDP were to increase the number or scope of people who are included in investigations, then this might have a negative impact not only on privacy but also on people's family lives (Samuel and Prainsack 2019b).

The search for new ethical boundaries

Considering these concerns, and in light of the coding versus non-coding DNA boundary being an unsuitable ethical and legal criterion in the context of FDP, several scholars have highlighted the importance of creating new categories and criteria to discern ethically acceptable uses of FDP from those which are more problematic. Some scholars suggest the use of reliability and validity thresholds. Reliability refers to consistency in the performance of the test across time and across places, whereas the test's validity captures the extent to which the test actually delivers on what it seeks out to test – in our case, the extent to which the markers for skin, hair, eye colour, biogeographical ancestry and age actually represent the category in question. One approach to finding a new ethical boundary could be to require that in real-life criminal investigations, only those tests that lie above a specific threshold of reliability and validity should be used. Some authors argue that only this would guarantee that there is a degree of certainty with which the phenotypic trait could be measured and conclusive meaningful results drawn in operative work (Smith and Urbas 2012; Murphy 2013; Samuel and Prainsack 2018a). The same scholars also purport that traits visible to the naked eye are less ethically sensitive to test for, and thus the visibility criterion acts as an additional useful ethical and legal boundary for acceptable FDP use (Smith and Urbas 2012; Murphy 2013). Seo et al. (2017), for example, argues that FDP should be used 'only for predicting features that are as perceptible as to human eyewitnesses by using only the markers that are completely free of any ethical disputes' (Seo et al. 2017). Proponents of this solution argue that this would bypass most, though not all, privacy issues. Despite the problems inherent in the comparison between human eyewitnesses and FDP, the visible versus non-visible distinction may indeed offer another useful criterion to assess the ethically acceptable use of FDP (Samuel and Prainsack 2018a).

Exposing implicit normativities in FDP

Exploring the ethical and social concerns arising from the practice of FDP and formulating appropriate new ethical boundaries as described in this chapter is commendable. But another step of ethical reflection is required to ensure that FDP is used in an ethically and socially responsible manner. Rather than merely ascribing ethical relevance to the physical realities of FDP – that is, to the DNA (i.e., coding versus non-coding), to the associated EVCs (how sensitive these externally visible traits are in terms of risks to privacy and of discrimination) or to the robustness of DNA analysis methodologies – we must analyse the implicit normativities (Cribb 2019: 1) that our very description of such physical and social realities implies. The aim of this process is not to abolish ethical uncertainty but to expose it, reflect on it and, if necessary, address it.

An important implicit normativity inherent in FDP is the assumption that prediction testing for appearance traits is less sensitive because these traits are supposedly less private. As described earlier, it is common among stakeholders and experts in the field to draw an ethical boundary between FDP testing for so-called 'externally visible' traits and all other traits that are clearly apparent to others. Eye colour, hair colour and skin colour, so the argument goes, are visible to the naked eye and thus not private to the person. Following this line of reasoning, age and biogeographical ancestry are considered more sensitive than externally visible traits because neither can be determined just by looking at a person. However, the meaning of 'private' here is not suitable for the digital era. In the pre-digital world, if somebody looked at a person's face, they could remember some features and 'store' these in their mind. These features, however, were not saved in a digital format in a database and made available for automated mining, as it is the case in today's practice. The digitisation and storage of externally visible features – such as by digitising custody images and other types of images (e.g., on social media) for police use – makes these features available to other organisations who can then, through their data, 'see' the person without the person being able to see them back. In other words, an ethical assessment that takes the external visibility of a trait as the main ethical boundary misses the ethically sensitive properties that emerge from the data system that this information is part of. While such asymmetry could be seen as justified in the context of a criminal investigation, it is nevertheless a power asymmetry that needs to be acknowledged as a property of the field of law enforcement and criminal justice, which are becoming increasingly digitised and datafied (e.g., Brayne 2017; Jefferson 2020). The structural and organisational design of DNA-based phenotyping for criminal investigation is set up in such a way that the data doubles of people are placed behind a metaphorical one-way mirror (Wellcome Trust 2016) where 'Big Brother and Company Man' (Kang et al. 2012) can see everything about the people, but the people cannot see anything about them. Here, the shift from 'old', STR-based DNA analysis to FDP is significant. The 'old' data doubles comprising STR-based DNA profiles could give information on *who* the likely perpetrator was but not *how* they were. FDP, in contrast, focuses on the how. In this manner, it could be argued that a larger part of the person is now placed behind the one-way-mirror than ever before, which makes the power asymmetry between data subjects and data users much more problematic than was the case in previous eras. What exacerbates this situation further are the increasing inroads that commercial companies are making into the development and deployment of security and law enforcement technologies. For example, the (scientifically unsound) practice of creating images of the projected facial features of offenders based on their DNA is pushed by commercial companies and already utilised for racist practices and human rights infringements in some parts of the world (Wee and Mozur 2019; Hopman 2020). In sum, the exploration of implicit normativities in this context exposes the ethical issues related to privacy concerns in the context of appearance inference and, specifically, reminds us that appearance prediction testing can lead to wider institutional and societal concerns relating to surveillance and power. As a result, it has highlighted the need to think of privacy as a collective societal interest as much as it is an individual one. By looking at the systems within which FDP is enacted, we see that risks to the privacy of individual people and specific groups always also affect larger institutional arrangements and practices.

The explicit normativity of solidarity

With these discussions in hand and the range of ethical and social uncertainties about FDP laid bare, we argue that the search for a new ethical boundary for FDP prediction testing would benefit from the inclusion of solidarity as a guiding principle for policy development. Technolegal practices currently focus on issues of validity, reliability, visibility etc. But as we described earlier, these very practices lose sight of the implicit normativities. We argue that taking a solidarity approach to ethics/technolegal practices can be useful here and, in the following sections, show how this can play out in practice

But, first of all, what is solidarity? In a nutshell, solidarity takes place when people accept 'costs' – emotional, financial, practical or otherwise – to support others with whom they recognise a similarity in a relevant respect. Such recognition of similarity is not the recognition of 'objectively' existing commonalities, such as having the same religion, gender or ethnic affiliation; we recognise as a similarity with others what we have learned to recognise as such. If we were told as children, for example, that people with a specific skin colour or religious affiliation or those who overeat or smoke are lazy or dangerous, then we are not likely to see these people as part of 'us'. If we were taught that all humans are similar in that we have weaknesses (some of us eat or drink too much; some of us smoke or gamble), then we are less likely to regard these people as different from us. In other words, solidarity does not neglect or deny differences between people, but it makes those characteristics and features constitutive for actions that *connect* people to each other. For example, solidaristic healthcare systems, such as the United Kingdom's National Health Service, are built on the understanding that all people are vulnerable to illness and need help in situations of weakness. Although some will incur much higher costs throughout their lives than others, such a healthcare system deliberately disregards this. As a result, it provides services according to need and not to people's ability to pay or to actual costs incurred (see also Prainsack and van Hoyweghen 2020). This is the core tenet of institutionalised solidarity.

While solidarity has been one of the key institutional features of continental European welfare states, it may not intuitively seem a helpful concept in criminal justice or for the use of FDP. Criminal investigation is a field of practice that explicitly seeks to 'sort out' those people who have committed crimes, so it relies on the very processes of 'othering' that solidarity overcomes; the criminal is the *other* who needs to be removed from the rest of society or rehabilitated to be able to re-enter it. This raises questions about whether, and indeed how, solidarity can be useful in helping us think through processes and policies in this field.

We argue that solidarity can have a role in shaping policies and technolegal practices in the criminal justice system. We illustrate this by first using the example of security before moving on to apply this example to a specific case of FDP use.

Solidarity-based approaches to security

As many authors have noted, many societies have experienced a process of securitisation in recent years and decades (e.g., Huysmans 2000; Balzacq 2010; Balzacq et al. 2016). This means that ever more social problems are reframed as problems of security; most recently, demographic and social transformations such as immigration have also been reframed as security issues. In other words, the way that security is understood has diversified to the extent that it now permeates virtually every other policy field: migration, finance and even environmental policies (through critical infrastructures protection: e.g., freshwater resources). Correspondingly, the targets of security have diversified as well; in the name of prevention, facilitated by data-intense policing, in theory, every citizen can become a target. This does not, however, apply to each citizen in the same manner; some of us attract more surveillance than others. As critical scholars have pointed out, those of lower socio-economic status are exposed to a much wider range of surveillance measures than others: surveillance in the form of data capture about their social and family activities and status, their movements across national and other borders etc. (Bridges 2017; Gilman and Green 2018; Milan and Treré 2020). Migrants with low socio-economic status are especially exposed to a much greater amount of surveillance than those with higher status (e.g., Heemsbergen and Daly 2017; Barenboim 2016). For the rich and privileged, it has become easier than ever to cross borders physically, to move money and assets across the globe, with fewer obstacles and often less visibility, especially in the context of financial transactions. It is the poor for whom it is becoming more difficult because they have been 'securitised' to a larger extent. This unequal distribution of data surveillance, particularly between those of different socio-economic status, is rooted in another implicit normativity that has bearing on FDP: some people's bodies and movements are seen as more dangerous than others. For FDP, this means that results pointing at members of such a 'securitised' group, such as migrants and other minorities, are likely to exacerbate the 'surveillance load' that is borne by that group for the sake of the security of the majority. We argue that a solidarity-informed understanding of security could, if not abolish, then mitigate this problem. Such a solidarity-informed understanding of security would recognise and be critical of notions of security which divide groups in discriminatory ways. Its starting point would be what all people have in common: that everyone wants a good life for themselves and their family, to move about freely, to live in good health and to be treated with respect. In other words, rather than othering minority groups from the rest of the population, a solidarity-based approach will draw together these populations (majority/minority) in terms of shared characteristics. Besides those general characteristics that we all have in common, there are also other, more specific ones that unite people across religious, language-related and 'cultural' divides. When a serious crime has been committed, the vast majority wants the victim to survive and recover, for their family to heal and for the perpetrator to be found.

Security as a public good, understood in such a way, is not as such in tension with individual rights and interests; the majority of the population across all societal groups has very similar interests in this regard: that the investigation is well done and effective, that it does not disrupt communities, that it does not pit societal groups against each other or increase stigmatisation and stereotyping. (The exceptions are the perpetrator and right-wing politicians.) Also, individual rights are not in tension with the public good of security here; that people's rights are respected and protected is a necessary condition for security to remain a public good. That individual freedoms of suspects are temporarily compromised is a necessary evil in criminal investigations; both the respect for individual rights and the respect for security as a public good, however, mandate that such compromises to individual freedoms are kept to a minimum.

In sum, instead of treating security as a public good and individual rights as competing or even conflicting interests (Pavone et al. 2016), a solidarity-based approach would see the protection of personal and group rights and interests as closely interlinked and as a necessary condition for societal security. In the remainder of this chapter, we will use the Netherlands case of Milica van Doorn, a young woman who was raped and murdered in 1992 (Toom 2010), to argue how forensic DNA testing could be used in such a way that it does not divide people and increase societal tensions but, in contrast, acknowledges and acts on what people have in common, thus enacting a notion of security underpinned by solidarity. We conclude by suggesting that such a solidarity-based understanding of security has specific implications for technolegal practices as they relate to the use of FDP.

How solidarity helped: FDP and the case of Milica van Doorn

In June 1992, a young Dutch woman, Milica van Doorn was found dead in a pond near the house where she lived. The 19-year-old had been out to a birthday party from which she never returned home. When her body was found the next morning, it was obvious that she had been raped and killed. Van Doorn's death had a lot of resonance in the entire country and received extensive media coverage (Toom 2010). Police thus faced even greater pressure to find the perpetrator than they normally would in such a tragic case. Unfortunately, a DNA profile developed from traces on Milica's body did not lead to a match with a profile in the national Dutch DNA database, which had been established in 1997. An eyewitness statement suggested a man with a Turkish appearance was seen on a bicycle near the place where Milica's body was later found, but no other leads meant that the case remained unsolved at this time. To close the case, police tested the DNA of nearly 300 individuals[5] in the years 2003 and 2005, but none of the men's DNA profiles matched the DNA profile obtained from the crime scene. Almost a decade later, in 2013, when biogeographical ancestry inference became a possibility, the crime scene DNA sample was tested for any such clues. It suggested that the suspected perpetrator's ancestry was from a geographic region including Turkey and neighbouring countries. This led police to suspect that the suspected perpetrator was of Turkish descent and the person on the bicycle could be involved. Police mapped the local area for people with Turkish ancestry and aimed to conduct a wider YSTR[6] dragnet of 7,000 men, specifically of Turkish descent, who had lived near the crime scene at the time of the murder. According to police officers involved in this case, due to political tensions between the Netherlands and Turkey at the time (around 2016), the decision was made to start DNA testing a smaller ring of 133 men. Familial searching – including dragnets based on YSTR DNA information – had been legalised in the Netherlands in 2012 (see M'charek et al. 2020). With the help of familial searching and other evidence, they were able to apprehend a suspect: the brother of one of the 133 men who had volunteered a DNA sample. The suspect himself had refused to volunteer a sample when initially asked. In 2018, the suspect was confirmed to be the perpetrator with the help of traditional STR-based DNA testing. He was sentenced to 20 years in prison, and the case of Milica van Doorn could be closed.

How does solidarity come into play here? What is remarkable in the Milica van Doorn case is that, although witness evidence and FDP testing indicated the involvement of a man belonging to an ethnic minority, it apparently did not increase social tensions and stigmatise minorities from Turkish backgrounds. This is even more remarkable given the structural and open racism prevalent in European countries, including the Netherlands (see also Van Oorschot and M'charek 2021). Police investigators were well aware of this, as well as of the need to proactively counteract stigmatisation and incitement to racist violence. Moreover, they were aware of the sensitivity of the situation: asking men who belonged to a stigmatised minority to provide a DNA sample which could potentially incriminate a relative.

In line with solidarity, which emphasises what people have in common and not what sets them apart, the police adopted an approach of collaborating with Dutch-Turkish communities to find the perpetrator, rather than 'othering' the Turkish population. Police officers who entered the homes of Turkish Dutchmen to obtain DNA samples, if they were not already familiar with them, were 'familiarised' with social and cultural norms prevalent in Dutch-Turkish communities (personal communication of the authors with people involved in the investigation, autumn 2019). 'Familiarising' here meant more than mere cultural sensitivity training for merely instrumental reasons – i.e., to get people to volunteer a DNA sample. It entailed an acknowledgement that the practices of others who may look different from what certain people know fulfil the same functions of the customs of the latter; it is just that the customs take different forms. By becoming familiar with people's practices in this way, they literally stop being 'the other', and they become similar to 'us'. This does not mean neglect or ignorance of difference but emphasis on similarities despite the existing differences. This, in turn, is exactly the spirit of solidarity, which signifies practices of mutual support between people who – despite all the things that separate them – act on what they have in common.

People involved in the investigation of the Milica van Doorn case told us that for the police officers taking the DNA from men from the Dutch-Turkish community, the experience of entering their homes and being welcomed with warmth and hospitality was very positive as well, which further strengthened a sense of solidarity in the community, across religious and cultural borders. 'The men who were approached were not offended', we were told by a Dutch police officer, 'but they were happy to help. They told us, "We have children, too!"' (personal communication of the authors with people involved in the investigation, autumn 2019). Out of all the men asked to volunteer a DNA sample, only two declined. (One of them later turned out to be the perpetrator.)

By giving this example, we do not mean to suggest that concerns about discrimination and stigmatisation of ethnic and other minorities in the context of FDP – or DNA profiling more broadly – are exaggerated. 'Othering' based on racialised and religious stereotypes doubtlessly happened in the Milica van Doorn case, including racist headlines in newspapers (discussed in Toom 2010; see also Van Oorschot and M'charek 2021). As noted, especially in societies where police and structural racism are widespread – and this certainly includes the Netherlands – the use of a technology that yields information about genetic proximity is inevitably at risk of serving discriminatory practices. What we want to demonstrate with this example, instead, is how an approach that does not play out the security of the supposed majority against the interests of a smaller group – however defined – can help mitigate the risk of discrimination. Ultimately, we argue, an understanding of security infused by solidarity is necessary to help ensure that FDP is used in a way that does not divide people from each other but emphasises what people have in common. Having children, deserving respect for one's home and one's custom and the desire to close the case were features that united people across different societal groups. For these commonalities to be visible and actionable, however, they need to be given room and attention.

Conclusion

This chapter started with an overview of how FDP is different from traditional forensic DNA profiling. The latter seeks to identify individuals whose DNA matches DNA profiles obtained from crime scene samples by comparing genetic markers located in (mostly) non-coding regions of the DNA, which do not disclose information about inner characteristics of people (e.g., personality traits, disease risks etc.). FDP, in contrast, has been designed for use in cases in which serious crimes have been committed and other investigatory methods have not yielded any leads on the perpetrator. In such cases, FDP can be used with the aim of inferring, with a particular statistical likelihood, what the suspected perpetrator may look like, including their biogeographical ancestry (i.e., what continental regions their biological ancestors came from), their appearance and their age. In order to infer these traits, FDP looks at both non-coding and coding regions of the genome. This means that a categorical distinction that has served as an ethical and sometimes legal boundary to separate acceptable from problematic uses of forensic DNA analysis previously – namely, the distinction between analyses of coding versus non-coding DNA – has lost its utility.

Searching for new criteria to assess using FDP in an ethical and socially responsible manner (if any, at all), we argue for a nuanced approach that does not merely replace the coding versus non-coding distinction with a new dichotomy (e.g., externally visible versus not externally visible traits). Such dichotomies fail to address the deeper ethical dilemmas underpinning the technological practices in question, such as the problem of structural racism in our societies, as well as the assumption that inferring characteristics that are visible to the naked eye is unproblematic in connection with privacy rights because of a narrow understanding of privacy that fails to see the systemic privacy risks emerging from the digitisation of policing. In order to address these deeper, structural issues, it is necessary to bring to light the 'implicit normativities' (Cribb 2019) inherent in the categories that we use to frame and discuss ethical issues.

We introduced solidarity as a particularly helpful concept to foreground systemic characteristics and collective practices while not losing individual needs and practices from the equation. As noted, solidarity is best understood as practices by which people support others with whom they consider themselves as similar in a specific way (e.g., due to a shared goal, a common thread or merely the recognition of others as fellow parents, human beings etc.). Policies and institutions informed by solidarity, in turn, take as their starting point what people have in common and not what sets them apart. An understanding of security informed by solidarity, we argue, is one that seeks to overcome fragmentation and diversity and to contribute to social cohesion and mutual support. Rather than being merely a programmatic statement, we believe that specific lessons can be derived from this for the use of FDP in practice: while broader questions need to first be asked about the permissibility/acceptability of FDP, when FDP is employed, it reminds us that what FDP results mean and how they will affect communities need to be established with and by members of these communities, jointly with those who carry out FDP investigations.

Notes

1 An example is the use of Y-STRs for biogeographic ancestry.
2 DNA is made up of four different bases: A, T, C and G.
3 DNA methylation is a process that can modify the activity of a specific DNA sequence without changing the sequence itself. It is an epigenetic mechanism that controls gene expression (which, in turn, is correlated with biological ageing).
4 For an overview, see www.visage-h2020.eu/.
5 This figure, as well as other aspects of the case description, was verified by some of the key investigators involved in this case (personal communication with the authors, Autumn 2019).
6 Y-STRs are short tandem repeats (STRs) on the Y-chromosome.

References

Ahuriri-Driscoll, A., Tauri, J. and Veth, J., 2021. Māori views of forensic DNA evidence: An instrument of justice or criminalizing technology? *New Genetics and Society* 40/3: 249–266.

Amelung, N. and Machado, H., 2021. Governing expectations of forensic innovations in society: The case of FDP in Germany. *New Genetics and Society* [online first].

Balzacq, T. ed., 2010. *Understanding Securitisation Theory: How Security Problems Emerge and Dissolve*. London and New York: Routledge.

Balzacq, T., Léonard, S. and Ruzicka, J., 2016. 'Securitization' revisited: Theory and cases. *International Relations* 30/4: 494–531.

Barenboim, D., 2016. The specter of surveillance: Navigating 'illegality' and indigeneity among Maya migrants in the San Francisco Bay Area. *PoLAR: Political and Legal Anthropology Review* 39/1:79–94.

Brayne, S., 2017. Big data surveillance: The case of policing. *American Sociological Review* 82/5: 977–1008.

Bridges, K. M., 2017. *The Poverty of Privacy Rights*. Stanford, CA: Stanford University Press.

Chow-White, P. A. and Duster, T., 2011. Do health and forensic DNA databases increase racial disparities? *PLoS Medicine* 8/10: e1001100.

Cino, J. G., 2017. Tackling technical debt: Managing advances in DNA technology that outpace the evolution of law. *American Criminal Law Review* 54: 373–421.

Cribb, A., 2019. Managing ethical uncertainty: Implicit normativity and the sociology of ethics. *Sociology of Health & Illness* n/a (n/a). doi: 10.1111/1467-9566.13010.

Dewey-Hagborg, H., 2015. Forensic DNA phenotyping in the news. *Bionoymous.me*. http://biononymous.me/forensic-dna-phenotyping-in-the-news/ (last accessed 23 October 2021).

Duster, T., 2005. Race and reification in science. *Science* 307/5712: 1050.

Enserink, M., 2011. Can this DNA sleuth help catch criminals? *Science* 331/6019: 838–840.

Gilman, M. and Green, R., 2018. The surveillance gap: The harms of extreme privacy and data marginalization. *NYU Review of Law & Social Change* 42: 253.

Granja, R. and Machado, H., 2020. Forensic DNA phenotyping and its politics of legitimation and contestation: Views of forensic geneticists in Europe. *Social Studies of Science* [online first].

Grewal, K., 2007. The 'Young Muslim Man'in Australian public discourse. *Transforming Cultures eJournal* 2/1.

Haga, S. B., 2006. Policy implications of defining race and more by genome profiling. *Genomics, Society and Policy* 2/1: 57–71.

Hagan, J., Levi, R. and Dinovitzer, R., 2008. The symbolic violence of the crime-immigration nexus: Migrant mythologies in the Americas. *Criminology & Public Policy* 7/1: 95–112.

Heemsbergen, L. and Daly, A., 2017. Leaking boats and borders: The virtu of surveilling Australia's refugee population. *Surveillance & Society* 15/3–4: 389–396.

Hindmarsh, R. and Prainsack, B., 2010. *Genetic Suspects*. Cambridge: Cambridge University Press.

Hopman, R., 2020. Opening up forensic DNA phenotyping: The logics of accuracy, commonality and valuing. *New Genetics and Society* 39/4: 424–440.

Huysmans, J., 2000. The European Union and the securitization of migration. *JCMS: Journal of Common Market Studies* 38/5: 751–777.

Jefferson, B., 2020. *Digitize and Punish: Racial Criminalization in the Digital Age*. Minneapolis, MN: University of Minnesota Press.

Kang, J., Shilton, K., Estrin, D., Burke, J. and Hansen, M., 2012. Self-surveillance privacy. *Iowa Law Review* 97: 809–847.

Kayser, M., 2015. Forensic DNA phenotyping: Predicting human appearance from crime scene material for investigative purposes. *Forensic Science International: Genetics* 18: 33–48.

Koops, B.-J., Prinsen, M. and Schellekens, M., 2006. Wanted: A tall blond dutchman—does the Netherlands set the stage in regulating forensic DNA phenotyping. *Tilburg Foreign Law Review* 13: 206.

Koops, B.-J. and Schellekens, M., 2008. Forensic DNA phenotyping: Regulatory issues. *Science and Technology Law Review* IX: 158–202.

M'charek, A., 2008. Silent witness, articulate collective: DNA evidence and the inference of visible traits. *Bioethics* 22/9: 519–528.

M'charek, A., 2020. Tentacular faces: Race and the return of the phenotype in forensic identification. *American Anthropologist* 122/2: 369–380.

M'charek, A., Schramm, K. and Skinner, D., 2014. Topologies of race: Doing territory, population and identity in Europe. *Science, Technology, & Human Values* 39/4: 468–487.

M'charek, A., Toom, V. and Jong, L., 2020. The trouble with race in forensic identification. *Science, Technology, & Human Values* 45/5: 804–828.

M'charek, A., Toom, V. and Prainsack, B., 2012. Bracketing off population does not advance ethical reflection on EVCs: A reply to Kayser and Schneider (letter to the editor). *Forensic Science International: Genetics* 6/1: e16–e17.

Milan, S. and Treré, E., 2020. The rise of the data poor: The COVID-19 pandemic seen from the margins. *Social Media+ Society* 6/3.

Murphy, E., 2013. Legal and ethical issues in forensic DNA phenotyping. *New York University Public Law and Legal Theory Working Papers Paper* 415.

Ossorio, P. N., 2006. About face: Forensic genetic testing for race and visible traits. *Journal of Law, Medicine & Ethics* 34/2: 277–292.

Pavone, V., Santiago Gomez, E. and Jaquet-Chifelle, D.-O., 2016. A systemic approach to security: Beyond the tradeoff between security and liberty. *Democracy and Security* 12/4: 225–246.

Prainsack, B. and van Hoyweghen, I., 2020. Shifting solidarities: Personalisation in insurance and medicine. In: Pulignano, V., Van Hoyweghen, I. and Meyers, G. (eds). *Shifting Solidarities. Trends and Developments in European Societies*. New York: Palgrave Macmillan. 127–151.

Ransohoff, K. J., Jaju, P. D., Tang, J. Y., Carbone, M., Leachman, S. and Sarin, K. Y., 2016. Familial skin cancer syndromes: Increased melanoma risk. *Journal of the American Academy of Dermatology* 74/3: 423–434.

Samuel, G. and Prainsack, B., 2018a. Forensic DNA phenotyping in Europe: Views "on the ground" from those who have a professional stake in the technology. *New Genetics and Society* 38/2: 119–141.

Samuel, G. and Prainsack, B., 2018b. The regulatory landscape of forensic DNA phenotyping in Europe. Project report. Available at: www.visage-h2020.eu/PDF/Deliverable_5.1_for_posting_online_DECEMBER_2018.pdf (last accessed: 23 October 2021).

Samuel, G. and Prainsack, B., 2019a. Civil society stakeholder views on forensic DNA phenotyping: Balancing risks and benefits. *FSI Genetics* 43: 102157.

Samuel, G. and Prainsack, B., 2019b. Societal, ethical, and regulatory dimensions of forensic DNA phenotyping. Project report. Available at: www.visage-h2020.eu/PDF/Delliverable_5.2_for_online_publication_vo1.pdf (last accessed: 23 October 2021).

Sankar, P., 2010. Forensic DNA phenotyping: Reinforcing race in law enforcement. In: Whitmarsh, I. and Jones, D. (eds). *What's the Use of Race?: Modern Governance and the Biology of Difference*. Cambridge, MA and London, England: MIT Press. 49–62.

Sankar, P., 2012. Forensic DNA phenotyping: Continuity and change in the history of race, genetics, and policiing. In: Wailoo, K., Nelson, A. and Lee, C. (eds). *Genetics and the Unsettled Past: The collision of DNA*. Piscataway: Rutgers University Press. 104–113.

Schneider, P. M., Prainsack, B. and Kayser, M., 2019. The use of forensic DNA phenotyping in predicting appearance and biogeographic ancestry. *Deutsches Ärzteblatt* 116: 873–880.

Seo, H. J., Cho, S., Lee, J. H., Lyoo, S. H., Kim, M-Y. and Lee, S. D., 2017. Forensic DNA phenotyping: A review in korean perspective. *Korean Journal of Legal Medicine* 41: 23–31.

Smith, M. and Urbas, G. F., 2012. Regulating new forms of forensic DNA profiling under Australian legislation: Familial matching and DNA phenotyping. *Australian Journal of Forensic Sciences* 44/1: 63–81.

Toom, V., 2010. Inquisitorial forensic DNA profiling in the Netherlands and the expansion of the forensic genetic body. In: Hindmarsh, R. and Prainsack, B. (eds). *Genetic Suspects: Global Governance of Forensic DNA Profiling and Databasing*. Cambridge: Cambridge University Press. 175–196.

Toom, V., 2012. Bodies of science and law: Forensic DNA profiling, biological bodies, and biopower. *Journal of Law and Society* 39/1: 150–166.

Toom, V., Wienroth, M., M'charek, A., Prainsack, B., Williams, R., Duster, T., Heinemann, T., Kruse, C., Machado, H. and Murphy, E., 2016. Approaching ethical, legal and social issues of emerging forensic DNA phenotyping (FDP) technologies comprehensively: Reply to 'Forensic DNA phenotyping: Predicting human appearance from crime scene material for investigative purposes' by Manfred Kayser. *Forensic Science International: Genetics* 22: e1–e4.

Van Oorschot, I. and M'charek, A., 2021. Un/doing race: On technology, individuals, and collectives in forensic practice. In: Hojer Bruun, M., Ayo Wahlberg, Rachel Douglas-Jones, Cathrine Hasse, Klaus Hoeyer, Dorthe Brogård Kristensen, and Brit Ross Winthereik, (eds). *The Handbook for Anthropology of Technology*. Singapore: Palgrave. 399–414.

Wallace, H. M., Jackson, A. R., Gruber, J. and Thibedeau, A. D., 2014. Forensic DNA databases—Ethical and legal standards: A global review. *Egyptian Journal of Forensic Sciences* 4/3: 57–63.

Wee, S.-L. and Mozur, P., 2019. China uses DNA to map faces, with help from the west. *New York Times*. Available at: www.nytimes.com/2019/12/03/business/china-dna-uighurs-xinjiang.html?action=click&module=Top%20Stories&pgtype=Homepage (last accessed 23 October 2021).

Weichert, T., 2017. Genetische Forensik und Datenschutz (Genetic forensics and data protection). Zeitschrift für Bürgerrechte und Gesellschaftspolitik. *Journal of Civil Rights and Social Policy* 218/2:123–134.

Wellcome Trust, 2016. *The One-Way Mirror: Public Attitudes to Commercial Access to Health Care*. Ipsos MORI Social Research Institute. Available at: www.ipsos.com/sites/default/files/publication/5200-03/sri-wellcome-trust-commercial-access-to-health-data.pdf (last accessed 23 October 2021).

Wienroth, M., 2018. Governing anticipatory technology practices. Forensic DNA phenotyping and the forensic genetics community in Europe. *New Genetics and Society* 37/2: 137–152.

Wienroth, M., Morling, N. and Williams, R., 2014. Technological innovations in forensic genetics: Social, legal and ethical aspects. *Recent Advances in DNA & Gene Sequences (Formerly Recent Patents on DNA & Gene Sequences)* 8/2: 98–103.

13

CONCEPTIONS OF CONSENT, FAMILY AND JURISDICTION IN FORENSIC GENETIC GENEALOGICAL SEARCHES

Erin Murphy

The announcement in April of 2018 that law enforcement officials in the United States had identified a serial rapist and murderer known as the Golden State Killer (GSK) heralded a new era in forensic genetic investigation. At a press conference announcing the arrest on the United States's official National DNA Day, investigators initially refused to disclose their methods. Instead, a litany of law enforcement officials, and even a victim's relative turned DNA advocate, spent time promoting the passage of a pending expansive DNA law and even repeatedly 'sham[ing]' a politician by name who had resisted his entreaties (Press Conference 2018, at 26:09–30:42). But within days, reporters revealed that the suspect had been traced through a crime scene profile uploaded to GEDMatch, a recreational genealogical database, which produced links to a dozen or so distant relatives (Selk 2019; Gafni 2018). After building out a family tree with 'thousands' of members, investigators honed in on Joseph James DeAngelo, who eventually pled guilty to the crimes (E. Murphy 2018, p. e6). The investigators who conducted most of the research became minor celebrities, and their success spawned a cottage industry of crime-solver television shows and new forensic genetic genealogical (FGG) service shops (Arnold 2020; Zhang 2019).

An FGG search differs from traditional forensic DNA in several respects. First, FGG searches involve a more invasive and comprehensive form of DNA testing than the traditional form of forensic genetic testing. FGG compares hundreds of thousands of single nucleotide polymorphisms (SNPs) in the coding region of the genome, rather than a couple of thousand short tandem repeat sections (STRs) deliberately chosen from non-coding or 'junk' regions of the DNA strand (Kling et al. 2021; Greytak et al. 2019; Kennett 2019, pp. 108–109). Second, FGG searches use largely unregulated commercial and recreational databases, rather than law enforcement databases subject to strict controls to safeguard of both privacy and quality assurance. Third, rather than look for an exact match, FGG searches

DOI: 10.4324/9780429322358-17

uncover scattered leads in a suspect's family tree, from which genealogists then reverse engineer the rest of the tree to isolate one branch or a single suspect (Kennett 2019). In this respect, FGG searches differ even from familial searches of DNA databases, which typically focus on close male relatives of a suspect and have a poor success rate at achieving even that. FGG searches also require law enforcement to comb through highly personal records – such as those containing birth, death, census, law enforcement, educational, financial, social media, real estate records and so on – to build out a family tree. In the course of the investigation, investigators might also seek DNA samples from biological relatives in order to help fill gaps or steer the direction of the investigation.

These complex FGG investigations also take place wholly in secret, and investigators in the US have deliberately hidden their true scope from the public. For instance, in the GSK case, later reporting revealed that neither investigators nor the DNA database companies had been forthcoming – and not just in the press conference, but in the year-long victory lap of media appearances that took place after the arrest (St. John 2020). In fact, it was not a hit in the GEDMatch database that broke the case. It was not only public databases that were searched. And DeAngelo was not the sole person who had been suspected of the crime and not even the only person from whom investigators had taken DNA. Instead, as explained later in this chapter, the investigation had involved dubious police practices, more genetic sleuthing and more innocent people placed under police suspicion than the police had let on.

Within two years of that event, hundreds of cases had used genetic genealogy methods to identify suspects, resulting in dozens of convictions (Kling et al. 2021, p. 2; Katsanis 2020, pp. 544–548 tbl. 2). Commercial providers like Parabon and Bode International have officially started offering FGG services, and other companies have joined suit (Katsanis 2020, p. 543). The searches have proven largely efficacious, relying on a combination of the size of the database searched, the discerning power of the search algorithm, the depth and range of coverage in non-genetic records necessary to build out family trees and the skill of the particular genealogist.

At the same time, there are some indications of increasing trepidation among the general public with regard to recreational genetics. Illumina, the company that dominates the recreational genetics-testing technology, 'hit a lull' after years of exponential growth; analysts have speculated that concerns about privacy may be a contributing factor, as well as possible exhaustion of the pool of interested consumers (Farr 2019). A recent announcement from the Pentagon in the United States may serve only to underscore this fear, as they warned service members of the risks of recreational genomics (Murphy & Zaveri 2019). And in the US, several legislators have spoken about enacting a total ban (Coleman 2020), and a few states have imposed restrictions on the practice, including the first comprehensive regulatory scheme in the state of Maryland (Ram et al. 2021).

The emergence of FGG affords a valuable platform from which to assess an incipient technolegal world – one in which cutting-edge technologies clash with well-established legal, social and political concepts. Forensic genealogy harnesses several distinct scientific advances, including the availability and affordability of inexpensive, large-scale single nucleotide polymorphism (or SNP) testing of the genome; the commercialisation of recreational genetics by private companies; and the development of open-platform sites to share genomic data (Kling et al. 2021, pp. 1–4). Hitched to one another by law enforcement, these technosciences pose a challenge to fundamental legal concepts that have long constrained law enforcement. Namely, they unsettle notions of consent, the spatial and relational limits on knowledge generation and conventional legal structures governing law enforcement access to information. Focusing primarily on the law and practice of forensic genetic genealogy in the United States, this chapter addresses each of these sites of conflict.

I Consent: What and who agrees?

The concept of consent resonates across a variety of legal domains. Consent carries legal force in commercial law, property law, health law, ethics, torts, criminal law and so on. Consent allows the law to shortcut difficult legal questions about rights or the superiority of claims by simply inquiring into a person's willingness to engage in or submit to a behavior or transaction. Indeed, there are relatively few areas over which the law claims the superior right to deny the power of adults to consent (Bergelson 2007, p. 175).

It is perhaps no surprise then that proponents of genetic genealogy often claim 'consent' as a legal authority for undertaking FGG investigations. But this invocation of consent as a legitimating mechanism raises questions: What is consent in this frame, and whose consent is important to legitimise FGG search methods? At a superficial level, genetic genealogy implicates two general strands of legal consent – those found in criminal procedure and in human subjects clinical research. But is the 'consent' invoked by advocates of genetic genealogy in fact recognisable as a conventional legal category, or is it instead a new form of consent that is legally unstable or contestable?

In US constitutional criminal procedure, virtually any law enforcement activity may be undertaken with the consent of the proper party, regardless of whether such methods would otherwise require formal legal authorisation. Consequently, 'consent' is an oft-cited justification for law enforcement actions, and courts focus more on the scope and breadth of valid consent (*Georgia v. Randolph* 2006) than on whether the police action is invalid notwithstanding consent. Critics have charged that courts' interpretation of consent belies a 'schism between the language of the governing standards, which emphasises freedom of voluntary choice, and actual case outcomes, where the Court repeatedly finds voluntary consent when reality would suggest otherwise' (Burke 2015, p. 522).

'Consent' in the medico-research context is a bit more challenging conceptually. At the most general level, US law – following international standards first embodied in the Declaration of Helsinki – has imposed detailed standards of 'informed consent' to protect human subjects of clinical experimentation (Couture 2004, p. 139). Generally speaking, a subject must be provided, in writing, with information detailing future risks under conditions that minimise coercion or undue influence (Brunts et al. 2012, p. 32). But 'advances in genetic and genomic research – in particular, the increasing emergence of large-scale population studies and genomic databases – have challenged traditional conceptions of informed consent' (McGuire & Beskow 2010, p. 362). McGuire and Beskow identify a number of ways in which biomedical genetic research defies conventional norms of consent. Because it is 'impossible to describe in detail or even to foresee all the future research for which [that specimen] might be used' or the 'future risks' of disclosing genetic material', (McGuire & Beskow 2010, p. 362), health ethics has embraced the notion of 'broad consent' as justifying genomic research in line with the spirit, if not the precise letter, of ethical informed consent. (45 C.F.R. § 46.116).

In short, in the clinical research context, consent ought to reflect a balance between the personal and privacy interests of genetic donors, who cannot feasibly be informed of all the consequences of wide-scale genomic research, and the public benefit of conducting such research. In the criminal justice context, consent ought to balance the values of individual autonomy and liberty and the goal of public safety. But whereas the clinical research community struggles to reconcile these divergent interests, the systems of law enforcement exhibit no such nuance. Instead, consent is invoked by law enforcement as implicitly authorising genetic genealogical searches or shunted off as an obligation binding only the private genealogical platforms.

Thus, for instance, in the wake of the Golden State Killer's arrest, much of the dialogue surrounding law enforcement's use of the GEDMatch database centred on the idea of 'consent'. And, in particular, on 'consent' as a commonplace concept, not tethered specifically to a legal definition, whether drawn from criminal justice, research, consumer contract, ethics or other relevant fields. Namely, both police and the public claimed that the persons who had chosen to upload and share their genomic data publicly had 'consented' to any and all uses of that information, and thus law enforcement should have as much access as any other party.

In fact, 'consent' seems an odd fit for an investigation undertaken in complete secrecy, in which the investigative methods and the duplicity of consumer sites had to be uncovered by journalists rather than disclosed openly by police. When law enforcement eventually admitted to using genetic genealogy to identify the Golden State Killer, the news reports repeatedly touted GEDMatch – an open-source, publicly available site – as the database used in the search (St. John 2019; Selk 2019). But in fact, the crime scene profile had been developed and uploaded first to FamilyTreeDNA, a closed consumer site (St. John 2020). After that site produced only remote leads, a private genealogist working on the case turned to MyHeritage and simply uploaded the profile there as though it were her own. It was that upload that, in fact, generated the critical link, not GEDMatch (St. John 2020).

The reason to dissemble is clear: unlike GEDMatch, where any member of the public could freely roam, both FamilyTreeDNA and MyHeritage were closed sites that reassured users that they would protect their privacy and, indeed, expressly stated that they would require legal process before releasing information to law enforcement (St. John 2020). FamilyTreeDNA had even signed on to a public declaration of 'best practices' for consumer genetics privacy that included refusing to make disclosures to police (Future of Privacy Forum 2018). Shortly afterward, a press account outed the company as having secretly granted the FBI access to its databases for over a year (Aldhous 2018; FamilyTreeDNA 2019b).

But even if genetic genealogical searches were, in fact, limited to open and public sites like GEDMatch, 'consent' still fits poorly as a framework. Prior to the news of the DeAngelo case, the GEDMatch terms of service warned that although the site was 'intended solely for genealogical research, we are unable to guarantee that users will not find other uses' (Kennett 2019, p. 112). Notwithstanding this alert, news that police had accessed the database incited blowback from some users, and within a month, GEDMatch had changed its policy to allow users to control whether law enforcement could access their data. But the terms whipsawed over a short period of time – toggling between default opt-in and default opt-out structures (Kling et al. 2021, p. 9) and, at one point, coming under criticism for defying its own stated policy disallowing searches for anything other than homicide or rapes (Aldhous 2019). Then, in late 2019, GEDMatch sold itself to a for-profit company called Verogen (Wong 2019; Aldhous 2019). But even that platform has been attacked for both inadequate security measures and the sudden reappearance of a large number of files that users had believed they had deleted (Kling et al. 2021, pp. 14 & n.59). Not every recreational genetics company is the same – not just in their policies, which vary widely (Kling et al. 2021, pp. 10–12), but in the extent to which the company has taken aggressive stances against law enforcement, including by challenging legal process in court, and by publishing transparency reports (Kling et al. 2021, p. 11; Katsanis 2020, p. 552).

In the full light of day, 'consent' is a talisman that lacks any discernible content, even as it operates conceptually as both a legitimating mechanism and a site of discourse over the practice of forensic genetic genealogy. Proponents are correct that the participants in a public and open database have voluntarily contributed their genomic data to that database, but they neglect to account for the misrepresentations and lack of candor on the part of law enforcement and the companies or the shifting sands that make any meaningful consent impossible. They also wholly neglect that most persons who participate in recreational sites are dimly aware of possible law enforcement access, falling far short of 'informed consent'. Participants chose to disclose data primarily motivated by a desire to connect with and find personal relationships; most have not contemplated the wide array of nefarious uses to which such data could be put (Murphy & Zaveri 2019; E. Murphy 2018, pp. e7–e8). And to the extent that participants on closed sites expressly did *not* consent to law enforcement access, law enforcement has largely escaped consequence for ignoring that lack of consent, whether in ethical or legal terms.

Perhaps more importantly, 'consent' is a poor proxy for the validity of this particular police technique, given that FGG implicates persons beyond the single consumer who submits a sample to a site. Indeed, that is its very purpose – to uncover and build out large networks of people, not to develop a better understanding of a single person's genomic identity. As clinical researchers have long observed, the 'informed consent' frame is ill suited to genomic research in this way. 'Broad consent' best addresses a sort of vertical or longitudinal informational and privacy interest of a participant in genomic research – it speaks primarily to unanticipated future uses of a research sample as regards the donor of that sample. But there is also a horizontal or latitudinal implication of genomics research, in that it pulls in hundreds or even thousands of other persons who may lack even basic awareness that their genomic profile, based on shared alleles, is under study, much less a voice in how or by whom that material is accessed.

To put it bluntly, to the extent that speaking of a GEDMatch or FamilyTreeDNA user's 'consent' to FGG searches by law enforcement already feels fictional, any 'consent' by their biological relations is wholly fantastical. Yet 'consent' remains the language with which the propriety of FGG searches is debated. One study by Verogen revealed that searches against the 'opt-in' profiles of the database led to equivalent success rates (roughly 80%) as searches against the 'opt-out' profiles (Kling et al. 2021, p. 9). Other studies have modeled databases and determined that a small-size database (say, of five million people) renders entire populations genetically transparent (Erlich et al. 2018, p. 690). In other words, the people who chose *not* to engage in consumer genetics remain fully visible to law enforcement because FGG can find them whether a particular relative 'opted in' or 'opted out'. Given what is known about the identification power of genetics, only a small fraction of the population need assent to recreational genetics (in any form) for the remainder of the population to be exposed to broad and unconstrained law enforcement access. But if that is to be justified as 'consent', it is a wholly new technolegal framework in operation, not any notion of 'consent' derived from existing fields.

II Family: Spatial and relational limits of knowledge

Genetic genealogical searches also create a second new technolegal world: that pertaining to concepts of 'family' or 'relatedness'. Typically, 'family' is conceived as a biological, legal and social construction. Most pertinently, families often coalesce around biology. But, equally importantly, they are also legally and socially structured. Legal arrangements like adoption, surrogacy and marriage form families in the absence of biological connectedness. Families carry legal benefits and burdens, affecting topics as diverse as taxation, property rights and immigration status (Heinemann et al. 2016). And even without official legal sanction, 'families' emerge from informal social connections that are understood and defined as describing bonds of a certain character: enduring, unbreakable and even involuntary.

Much has been written about the post-modern family and the particular intersection of genetics, biology and our social and legal notions of relatedness, particularly as regards family law and public law. Some of this work has observed the disconnect between the biological family and the social one and the ways in which the law can choose to privilege one concept of family over another (NeJaime 2017, pp. 2017–2018). But less attention has been focused on the way in which the 'family' is deployed in the criminal justice context, especially in light of the increased use of forensic genetics to identify suspects. The one exception is perhaps with respect to the practice of familial searching (FS). But as noted earlier, FS differs from FGG in several respects. Most pertinently, FS relies on the conventional set of forensic markers to conduct the search, which means they have lesser implications for privacy and a generally less intrusive scope – although dragnet searches using the Y chromosome may more closely resemble FGG methods (M'charek et al. 2020). Second, familial searches rely on convicted offender or arrestee databases – databases that are closely regulated, in terms of both privacy and quality control, rather than voluntary public databases that lack any verification, quality-control or access restrictions.

FGG upsets conventional social and legal ideas of 'family' in several ways. Like FS, FGG biologises social categories, thus blurring the line between biological and social meaning (M'charek et al. 2020, p. 809). In earlier work, I observed that FS promotes the notion of certain families as criminogenic, tacitly planting the seed of a biological predisposition for deviance or criminal behavior (Murphy 2015, pp. 204–205). FS also has the effect of isolating only the relatives of convicted and arrested persons for special scrutiny, notwithstanding any indication that the perpetrator of the offence was, in fact, related to a databased person (Murphy 2015, p. 207). Also, because disparate law enforcement policies have led to disparate rates of arrest and conviction for different demographic groups in the United States, FS also inherently targets the innocent relatives of poor people of colour at disproportionate rates.

FGG raises similar concerns about 'families' as traditional FS methods, but FGG avoids some of the most pernicious aspects of FS while also raising fresh questions.

First, for instance: because FGG is conducted in publicly sourced databases, rather than compulsory offender databases, FGG avoids the criticism that FS methods unjustly target only certain families. Put simply, the problematic notion that FS is justified because 'crime runs in families' is largely absent in FGG since commercial databases contain a random assortment of voluntary contributors linked by nothing more than a common interest in recreational genetics.

At the same time, the process of creating sprawling family trees has the power to expose sensitive information about families, even without conducting a single genetic test. An investigator who links a thousand people through biology, who otherwise have no connection in the world, may discover in those birth and death and educational and financial and marriage and social media records a great deal about a family's predispositions or proclivities. Patterns could emerge that reveal religious or political affiliations or propensities toward miscarriage, infertility, chronic illness, early demise, addictions or even likely success or failure in marriage and education. Such findings may not be genetic in the sense that they are divulged through medical testing, but they are genealogical in the sense that they derive observationally from surveying the records of a cohort of people that are linked primarily through biology, rather than social association. In short, genetic genealogy creates a biological family at scales previously unknown, displacing and divorced from the conventional social or legal family.

FGG does probe families genetically; it harnesses phenotypic DNA testing techniques to reveal expressed physical traits (Kennett 2019, p. 15; Greytak et al. 2019, p. 107). Although typically, such traits are considered harmless, externally visible characteristics unlikely to implicate privacy, such as hair colour, eye colour or skin tone, trait testing opens the door to surveillance based on group attributes and exacerbates threat to family privacy. In the GSK case, for instance, the genetic genealogist advised law enforcement, based on DNA test results, that the suspect 'has a fair amount of Italian in him', leading the team to alert on the branch named 'DeAngelo' (Gafni 2018), and that the suspect probably had blue eyes and had prematurely balded (H. Murphy 2018), causing investigators to isolate one of six family members who met that description. One of the two false leads in the investigation arose because of similarities in a 'rare genetic marker' found in both the crime scene DNA sample and the elderly 'suspect' – a trait also apparently used to exclude members from the law enforcement investigative team, due to concerns that the perpetrator had a law enforcement background (Barnes 2018; Balsamini 2018). Surveillance of families based on presumed ancestry, physical traits or the presence or absence of certain genetic markers also runs the risk of exposing previously unknown offspring or incidences of nonpaternity or of upsetting long-held understandings of heritage or group belonging (Copeland 2020).

Second, compared to FS, FGG searches are arguably less subject to criticisms of racial bias or targeting. Recreational genetics databases at present contain the profiles of persons of European descent in higher proportions than those from other groups (Erlich et al. 2018, p. 690). Consequently, while FS methods are subject to criticism for perpetuating racial inequities, FGG searches target racial and ethnic groups not historically associated with police abuse.

Database composition may also diminish concern that genetic associations will bleed into an assessment of group identity: the fear that FGG will sow the seeds that white people are biologically prone to crime is simply less acute than it would be in the case of a less politically or economically powerful demographic group. One researcher has drawn a direct comparison between the pressure placed by consumers on a California recreational genetics company to 'start research on molecular determinants of sexual orientation' (Jabloner 2019, p. 2) and the state's forensic database. The resonance of biological determinism simply cannot be decontextualised from the group that is claiming it: whereas, for the predominantly white and affluent participants in recreational genetics, a 'gay gene' could be viewed as freeing them from arguments that orientation is choice rather than destiny, for marginalised or oppressed groups, genetic determination may condemn and stigmatise rather than liberate (Jabloner 2019, pp. 12–14).

That said, FGG hardly eradicates concerns of racial inequity. There is some indication that FGG database composition incentivises law enforcement to focus on the relatively small number of crimes involving white female victims, as opposed to the proportionately larger number of black male victims (Stern & Zhang 2021).

Third, and most troubling, the long reach of FGG – linking together persons through biology who otherwise have no social connections – upsets the whole notion of 'family' and 'relatives' in both the social and the legal sense. In the GSK investigation, genealogists constructed a 25-branch family tree with thousands of members (E. Murphy 2018, p. e6). Were a newspaper to print that tree, how would a typical member of the public respond? Would the third and fourth cousins of the suspect be considered 'relatives' of the GSK in any meaningful social sense? Would the taint of the GSK's actions cast a shadow over his biological relations, or would the social distance between branches sever one limb from another? It has been observed in a related context that the legal system can be particularly insensitive to the effect on the DNA donor of using samples provided for one purpose in order to achieve a different purpose, a phenomenon labeled 'decontextualized families' (Haimes & Toom 2014, p. 286).

In this respect, FGG challenges and alters our understanding of the significance of a 'family'. Whereas we might worry about treating a son as his father's 'genetic informant', it is less worrisome that a fourth cousin ends up being the starting point of an investigation. Indeed, in most cases, the fourth cousin may never learn of the connection – it is likely that the relatives of the GSK in the database still to this day do not realise that it was their DNA profiles that ultimately led investigators down the path that ended in GSK's arrest. Yet these troves of data exist. In a small community, it might take only a handful of investigations before law enforcement has fully mapped out the connectedness of a significant portion of its community. Those trees might remain in storage on police computers or, worse, on the unsecured home computers of genealogists – perhaps with the notes exposing mis-paternity events or secret liaisons – thereby linking or disconnecting people in the eyes of law enforcement in ways not evident in the social world.

In this way, whether the leads consider themselves 'family' or know of their role in the investigation, the long reach of FGG investigations enables invasions of privacy not just of 'suspects', but of all members of a biological tree. Again, in the DeAngelo case, investigators followed two leads all the way through to a genomic sample, but neither panned out. Specifically, investigators subpoenaed the identity and information of a consumer who had purchased and uploaded a kit to FamilyTreeDNA, but it turned out that the buyer was a woman who had uploaded her elderly father's DNA for recreational research; although law enforcement obtained a warrant to take DNA from him in his nursing home, he voluntarily submitted a sample, but the lead proved useless (Aldhous 2018). The FBI also approached a woman whose brother they considered the prime suspect; the sample cleared the brother but pointed investigators further down the family tree on the fourth site they accessed, Ancestry.com (St. John 2020).

Surveillance of innocent people, not just criminal suspects, is hardwired into the FGG process: the whole premise is that investigators comb through genetic and non-genetic databases to create family trees that reflect biological, rather than social, selves that lead them to the perpetrator (Kling et al. 2021 p. 7; Greytak et al. 2019, pp. 107–108, 111). These biological branches are thus brought unwittingly into law enforcement surveillance, even if they are not themselves suspected of the crime – and even if no one in their known social 'family' is. Tellingly, one family member wrongly drawn under suspicion in the Golden State Killer investigation refused to divulge their identity, stating that 'she did not want the family's name publicly linked to the case' and making a point to note that police 'were able to rule out people in my dad's [family] tree'. (Balsamini 2018).

Of course, some people may delight in helping solve a case or take pleasure in a personal connection to an important investigation. But even in the case of a willing participant in the investigation, there arises the possibility of abuse. In the United States, there has been at least one case in which investigators sought a voluntary sample from a distant relative in the 'tree' of interest. Having learned about this personal connection to the case, that relative – who turned out to be a genealogical hobbyist – posted the data in an online community in an effort to 'solve the crime' by crowdsourcing the suspect's identity. In jurisdictions without adequate laws governing the collection, sharing or testing of another person's genetic material – like most jurisdictions in the United States – such an investigation may also end in rogue crime enthusiasts engaging in acts of dubious morality or questionable ethics.

III Jurisdiction: Information and investigative governance

Genetic genealogy opens a new technolegal frontier in terms of jurisdiction and governance. The field of genetics is intensely intersectional; it crosses numerous borders. First, genetics transcends *physical* borders: biological connections cross national and international boundaries. Second, genetics transcends *temporal* boundaries; a single person's genetic information exposes the genetic information of their biological relatives, whether living, dead or yet to be born. Third, genetics defies *disciplinary* bounds as the lines that separate forensic, recreational, research and clinical uses of genetics blur.

Yet legal and ethical governance structures remain intensely localised within one spatial, temporal and disciplinary bound. Even regimes that govern populations as large and diverse as the EU or the United States are nonetheless often constrained within political, temporal and disciplinary boundaries. To begin, criminal justice regulation is notoriously localised, as evidenced by the struggle to manage cross-jurisdictional offenses and investigations in the face of interstitial, incomplete or conflicting rules (Ghappour 2017). Even treaties for the exchange of information or persons or cooperative agreements like the EU Prum Decisions, may smooth national differences but don't override them, and those endeavours have met with significant implementation roadblocks (Toom et al. 2019).

FGG presents a particular refinement of this problem. It is both intensely localised and broadly diffuse. A single person in a single jurisdiction may upload a DNA profile, but the connections made from that profile and the data inferable from it may cross national borders. Even if a particular 'search' is local, the data returned may profoundly affect the privacy interests of a broad number of people across a wide range of jurisdictions. To the extent that different legal regimes might offer greater or lesser protection to genomic data, whether used by law enforcement or ordinary persons, FGG methods can effectively nullify those choices.

Disciplinary boundaries are also ineffective constraints. A jurisdiction may circumscribe law enforcement collection and use of one type of DNA system, but that simply encourages law enforcement to turn to other systems or other jurisdictions that operate more permissively. Thus, for instance, a jurisdiction may have strict laws regulating compulsory DNA collection and searches in forensic database of convicted offenders. But stymied investigators can simply look to a neighboring jurisdiction with less restrictive rules (defying spatial constraints) or to DNA repositories outside that governance structure, such as public recreational sites (defying disciplinary constraints). For example, the GSK investigation utilised a law enforcement forensic database, a research-oriented Y-STR database, and both open and closed recreational sites (St. John 2020). Police can also leverage their power to force entities to divulge information. In the United States, courts have ordered GEDMatch, FamilyTreeDNA and a Sorenson molecular genealogy website to divulge otherwise private information (Aldhous 2018; St. John 2019). Soft power has also worked: the operator of FamilyTreeDNA, when asked about why he had surreptitiously granted law enforcement regular access to the site's genetic data, first replied that the police had essentially threatened him, telling him that 'if I didn't find a way to work with him, I would perpetually be dealing with a subpoena' (Marcus 2019). In another case, law enforcement simply reclassified an offence as a sexual assault in order to comply with the site's terms of access, then later recharged it as the actual offence – a burglary, which would not have qualified – once the target was identified (St. John 2019).

In addition to the incentive for law enforcement to expoloit the lack of firm spatial, disciplinary and temporal borders to evade legal regulation, the nature of genomic data makes it likely that even investigations conducted according to the rules of one regulatory structure invariably implicate interests outside that structure. By way of example, imagine a user in the United States who voluntarily submits a DNA profile to an open-access recreational site for the purposes of making familial connections. Unbeknownst to that person, law enforcement investigators are searching the site with a crime scene sample and discover that the person is a fourth-degree relative of the suspect. As investigators begin to build out that person's family tree, both the 'branches' and the investigation necessary to map them may easily end up crossing and recrossing national borders – an aunt who moved abroad but a child of that aunt who then moved back or a child who moved abroad and became a foreign citizen. The possibilities are many: the point is that the investigation, and those drawn into it by virtue of nothing more than their biological relatedness, may technically stay 'within' a legal jurisdiction while revealing information and prompting investigation of those theoretically outside its reach.

Which of several conflicting legal regimes should govern those implicated parties? Suppose, for instance, that one nation bars genetic genealogical searches or restricts them under tight conditions. For instance, the European General Data Protection Regulation imposes stringent rules on the use of data for a purpose other than the one for which it was collected, even as most US jurisdictions have no such restriction (Pormeister 2018, pp. 707–708). Can an investigation conducted entirely within the United States, consistently with US law, involve discovery of data pertaining to foreign citizens that would breach the laws governing those persons? Shortly after the news broke that FamilyTreeDNA had been cooperating with law enforcement in the United States, the company automatically opted out all customers who had previously created accounts that self-identified as being from the EU, although later-created accounts were presumptively opted in. FamilyTreeDNA subsequently announced its intent to comply with the Privacy Shield Framework designed by the US Department of Commerce (FamilyTreeDNA 2019a; Privacy Shield Framework 2021), but then European authorities called that framework into question (Case C 311/18; FDPIC 2020).

These conflicts are not limited to geographic boundaries alone: imagine, for instance, that investigators have cause to believe that access to a particular medical DNA repository may further the investigation. Should law enforcement rules govern searches in genetic repositories collected for health purposes, or should health rules govern? And what qualifications or standards should govern the searchers? Medical or research ethic regimes typically do not apply to law enforcement actors, and many genealogists are amateurs-turned-professional who lack any formal certification process or ethical training. Moreover, the multiple purposes for which DNA is collected, and the many repositories in which it is stored, present a challenge even for domestic investigations. In the United States, a complex patchwork of laws govern the collection, testing and storage of biological material (University of Minnesota 2020). And historically, scientific or medical research is regulated by the field of ethics, rather than law (Pormeister 2018, p. 707). Genomic data may be gathered for reasons as divergent as employment, insurance, criminal investigation, patents, research and clinical or medical examinations. Perhaps the only thing that can be said about the 'law' governing genomics is that 'the law underlying genomics is currently unclear, poorly understood, and contested' (University of Minnesota 2020).

These questions are difficult enough on their own terms without adding the further complication that samples collected under one set of conditions can end up used for something else. In one prominent case involving a Y-STR search in the United States, a man was wrongly implicated in a rape-homicide as a result of an anonymous sample donated by his father in connection with a health research project (Murphy 2015, pp. 201–203). Given that genomic data can never truly be anonymised (Edge 2017, p. 5671; Erlich 2018, pp. 691–692) and that it can be and already has been transferred as an asset in sales (Kling et al. 2021, p. 13), genetic information may readily detach from the original conditions under which it was collected. Even limiting access to only certain portions of genomic data cannot safeguard against misappropriation, as studies have shown that researchers can extrapolate the remainder of the genome reliably from limited portions (Edge & Coop 2020; Edge 2017, p. 5671).

Lastly, once genetic data is exposed, even a little bit, it becomes nearly impossible to claw back the information. Genetics reach across time, not just distance, reproduced in the chain of descendants in perpetuity. Lines may come to an end, and mutations and alterations may create subtle variations, but the capacity of genetic information to connect persons forward and backward through history remains strong (Foster et al. 1998). Thus, the genetic privacy implicated by police investigations stretches far beyond a single target or even the target's immediate relatives. Collection and use of genetic data changes meaning as the science and technology of genomics itself evolves, even as the legal regimes may be captured in a particular moment in time. In this way, even a limited amount of genomic information shared today by a single donor implicates not just that person, and not just limited information about that person, but the genomes of their descendants.

In sum, protection of genetic data becomes a question that extends far beyond a single donor, in a single legal regime, at a single moment in time and for a single purpose. Rather, FGG constructs a new technolegal world in which effective governance must sweep backwards and forwards in time, across enormous numbers of persons, through state and national borders and to data collected not just for criminal justice purposes but for recreational, research and health reasons. FGG presents a challenge to the basic precept of democratic legitimacy: the belief that a practice can be constricted and regulated a priori. And it generates a new legal reality in which a narrow decision to share genomic data, made by one person at one moment in time for one specific reason, can reverberate across time and space, affecting countless others.

IV Conclusion

In many respects, FGG is simply the culmination of a journey begun decades ago, when forensic genetic databases first took root. Although initially viewed as repositories of 'the usual suspects', forensic DNA databases quickly bloomed into all-purpose means of identification. The US Supreme Court has even implicitly suggested that the harmless DNA 'fingerprint' might one day be as commonplace a biometric identifier as its physical namesake (Murphy 2015, pp. 157–158).

But DNA is no ordinary 'fingerprint', and its particular qualities mean that the legal and ethical structures that have previously served to constrain police practices no longer operate as effectively. As this chapter explains, conventions of 'consent', 'family' and even 'law' or 'oversight' falter when put to the test by a technology with such capacious ability to identify and reveal.

Genetic data is not shared by 'consent' when one of the parties to the agreement withholds the truth, misrepresents their purpose or changes the terms after the fact without notice or repercussion. There can be no 'consent' without a cognisable scope or limit to the information shared or when the consent is elicited from one person whose single decision overrides the individual rights of countless others.

Similarly, to speak of 'families' or 'family trees' in the context to genetic genealogy is an effort to sanitized a much more sinister reality. The sprawling branches compiled by law enforcement, reaching back decades and centuries, across countries and continents and forward through generations, bear no relation to 'family' as we inhabit it in a lived and vibrant form of social connection and identity. A fourth cousin twice removed is not a 'family member' who participates in genetic genealogy; they are a distant biologically related stranger whose existence has been stitched to another's in service of penological objectives. And the secrets or privacies of life revealed in the process now sit on insecure hard drives, not within a cherished family bible or oral history – or, better still, lost altogether in time.

And lastly, to the extent that some notion of legal or regulatory governance may apply to circumscribe such searches, these efforts are readily confounded by the divergent regimes, disciplines and legal structures that superintend the fields implicated by genetic genealogy. No single governance regime easily confines the use of genetic genealogy, which suggests either the emergence of cooperative regulatory structures or the abdication of any hope of governance at all.

The emerging technolegal world of genetic genealogy is one that transcends physical, spatial, disciplines and temporal boundaries. It enables surveillance and evasion of regulatory control by operationalising across time and space, cabined by principles of biology, rather than social or legal ordering. Although debates around genetic genealogy tend to summon familiar concepts like consent, family and governance, closer examination of those terms reveals that they falter in the context of this potent new investigative method. In their place, we must devise new language and new concepts to describe and understand these genetic networks and to regulate and monitor the technolegal world they enable.

References

45 C.F.R. § 46.116. Available at: www.law.cornell.edu/cfr/text/45/46.116. (Accessed: 13 May 2021).

Aldhous, P. (2018) 'Cops forced a company to share a customer's identity for the golden state killer investigation', *Buzz Feed News*, 1 May. Available at: www.buzzfeednews.com/article/peteraldhous/family-tree-dna-subpoena-golden-state-killer (Accessed: 12 May 2021).

Aldhous, P. (2019) 'The genealogy website that helped crack the golden state killer case has been bought by a forensic genetics firm', *BuzzFeed News*, 9 December. Available at: www.buzzfeednews.com/article/peteraldhous/the-genealogy-website-that-helped-crack-the-golden-state (Accessed: 12 May 2021).

Arnold, C. (2020) 'The controversial company using DNA to sketch the faces of criminals', *Nature*, 585, 178–181, 9 September. Available at: www.nature.com/articles/d41586-020-02545-5 (Accessed: 12 May 2021).

Balsamini, D. (2018) 'Cops took DNA from innocent man in nursing home in search for serial killer', *New York Post*, 28 April. Available at: https://nypost.com/2018/04/28/cops-took-dna-from-innocent-man-in-nursing-home-in-search-for-serial-killer/ (Accessed: 12 May 2021).

Barnes, T. (2018) 'Golden state killer: Police using genealogy website wrongly identified innocent man in nursing home as suspect', *Independent* (U.K.), 28 April. Available at: www.independent.co.uk/news/world/americas/golden-state-killer-2018-case-solved-identity-genetic-genealogy-websites-dna-joseph-james-deangelo-arrest-trial-court-a8326946.html (Accessed: 12 May 2021).

Bergelson, V. (2007) 'The right to be hurt: Testing the boundaries of consent', *George Washington Law Review* (25), pp. 165–236.

Brunts, E.M., Hermes, C.D. & Rusczek, A.P. (2012) 'The international clinical trials roadmap: Steering clear of legal and practical roadblocks', *Journal of Health & Life Sciences Law* 5.

Burke, A.S. (2015) 'Consent searches and fourth amendment reasonableness', *Florida Law Review* 67, pp. 509–563.

Case C 311/18, Data Protection Commission v. Facebook Ireland, Ltd. [2020] ECLI:EU:C:2020:559 (16 July 2020). Available at: https://curia.europa.eu/juris/document/document.jsf?text=&docid=228677&pageIndex=0&doclang=en&mode=lst&dir=&occ=first&part=1&cid=9791227.

Coleman, E. (2020) 'One state may become the first to ban law enforcement use of genealogy databases', *Route Fifty*, 21 January. Available at: www.route-fifty.com/public-safety/2020/01/utah-dna-databases/162544/ (Accessed 13 May 2021).

Copeland, L. (2020) *The Lost Family: How DNA Testing is Upending Who We Are*. New York: Abrams Press.

Couture, J.J. (2004) 'The changes in informed consent in experimental procedures: The evolution of a concept', *Journal of Health & Biomedical Law* 1.

Edge, M.D. et al. (2017) 'Linkage disequilibrium matches forensic genetic records to disjoint genomic markers', *PNAS* 114(22), pp. 5761–5776.

Edge, M.D. & Coop, G. (2020) 'Attacks on genetic privacy via uploads to genealogical databases', *eLife* 9, p. e51810. doi: 10.7554/eLife.51810.

Erlich, Y. et al. (2018) 'Identity inference of genomic data using long range familial searches', *Science* 362, pp. 690–694, 9 November. doi: 0.1126/science.aau4832.

FamilyTreeDNA, Privacy Statement (2019a) Available at: www.familytreedna.com/legal/privacy-statement (Accessed: 24 January 2020).

FamilyTreeDNA. (2019b) 'Press release: Connecting families and saving lives, FamilyTreeDNA press release', Available at https://blog.familytreedna.com/press-release-connecting-families-and-saving-lives/ (Accessed: 12 May 2021).

Farr, C. (2019) 'Consumer DNA testing has hit a lull—here's how it could capture the next wave of users', *CNBC*, 25 August. Available at: www.cnbc.com/2019/08/25/dna-tests-from-companies-like-23andme-ancestry-see-sales-slowdown.html (Accessed: 24 January 2020).

FDPIC, Federal Data Protection and Information Commissioner (2020) *FDPIC Considers CH-US Privacy Shield Does Not Provide Adequate Levels of Data Protection*. FDPIC. Available at: www.edoeb.admin.ch/edoeb/en/home/latest-news/media/medienmitteilungen.msg-id-80318.html (Accessed: 12 May 2021).

Foster, E.A., Jobling M.A., Taylor, P.G., Donnelly, P., deKnijff, P., Mieremet, R., Zerjal, T. & Tyler-Smith, C. (1998) 'Jefferson fathered slave's last child', *Nature*, 5 November.

Future of Privacy Forum (2018) 'Privacy best practices for consumer genetic testing services', 31 July. Available at: https://fpf.org/blog/privacy-best-practices-for-consumer-genetic-testing-services/ (Accessed: 13 May 2021).

Gafni, M. (2018) 'Exclusive: The woman behind the scenes who helped capture the golden state killer', *San Jose Mercury-News*, 27 August. Available at: www.mercurynews.com/2018/08/24/exclusive-the-woman-behind-the-scenes-who-helped-capture-the-golden-state-killer/ (Accessed: 12 May 2021).

Georgia v. Randolph, 547 U.S. 103, 106 (2006).

Ghappour, A. (2017) 'Searching places unknown: Law enforcement jurisdiction on the dark web', *Stanford Law Review*, pp. 1075–1136.

Greytak, E.M., Moore, C. & Armentrout, S.L. (2019) 'Genetic genealogy for cold case and active investigations', *Forensic Science International* 299, pp. 103–113.

Haimes, E. & Toom, V. (2014) Hidden in full sight: Kinship, science and the law in the aftermath of the Srebrenica genocide', *New Genetics and Society* 33(3), pp. 277–294. doi: 10.1080/14636778.2014.946004.

Heinemann, T., Helén, I., Lemke, T., Naue, U. & Weiss, M. (2016) *Suspect Families: DNA Analysis, Family Reunification and Immigration Policies*. Surrey: Routledge.

Jabloner, A. (2019) 'A tale of two molecular Californias', *Science as Culture* 28(1), pp. 1–24. doi: 10.1080/09505431.2018.1524863.

Katsanis S.H. (2020) 'Pedigrees and perpetrators: Uses ofDNA and genealogy in forensic investigations', *Annual Review of Genomics and Human Genetics* 21, pp. 535–564, Table 2. Available at: www.annualreviews.org/doi/full/10.1146/annurev-genom-111819-084213#_i3 (Accessed: 12 May 2021).

Kennett, D. (2019) 'Using genetic genealogy databases in missing persons cases and to develop suspect leads in violent crimes', *Forensic Science International: Genetics* 301, pp. 107–117. https://doi.org/10.1016/j.forsciint.2019.05.01

Kling, D., Phillips, C., Kennett, D. & Tillmar, A. (2021) 'Investigative genetic genealogy: Current methods, knowledge and practice', *Forensic Science International: Genetics* 52. Available at: https://doi.org/10.1016/j.fsigen.2021.102474 (Accessed: 12 May 2021).

M'charek, A., Toom, V. & Jong, L. (2020) 'The trouble with race in forensic identification', *Science, Technology & Human Values* 45(5), pp. 804–828. doi: 10.1177/0162243919899467.

Marcus, A. (2019) 'Customers handed over their DNA. The company let the FBI take a look', *Wall Street Journal*, 22 August.

McGuire A.L. & Beskow L.M. (2010) 'Informed consent in genomics and genetic research', *Annual Review of Genomics and Human Genetics* 11, pp. 361–381. doi:10.1146/annurev-genom-082509-141711.

Murphy, E. (2018) 'Law and policy oversight of familial searches in recreational genealogy databases', *Forensic Science International* 292, pp. e5–e9.

Murphy, E.E. (2015) *Inside the Cell: The Dark Side of Forensic DNA*. New York: Nation Books.

Murphy, H. (2018) 'She helped crack the Golden State Killer case. Here's what she's going to do next', *New York Times*, 29 August. Available at: www.nytimes.com/2018/08/29/science/barbara-rae-venter-gsk.html (Accessed: 12 May 2021).

Murphy, H. & Zaveri, M. (2019) 'Pentagon warns military personnel against at-home DNA tests', *New York Times*, 24 December. Available at: www.nytimes.com/2019/12/24/us/military-dna-tests.html (Accessed: 24 January 2020).

NeJaime, D. (2017) 'The nature of parenthood', *Yale Law Journal* 126, pp. 2260–2376.

Pormeister, K. (2018) 'Genetic research and applicable law: The intra-EU conflict of laws as a regulatory challenge to cross-border genetic research', *Journal Law & Bioscience* 5(3), pp. 706–723.

Press Conference (2018) 'Golden State Killer', 25 April. Available at: www.youtube.com/watch?v=V5umyT0JH3s (Accessed: 12 May 2021).

Privacy Shield Framework (2021) Available at: www.privacyshield.gov/Program-Overview.

Ram, N., Murphy, E.E. & Suter, S.M. (2021) 'Regulating forensic genetic genealogy', *Science*, 24 September.

Selk, A. (2019) 'The ingenious and "dystopian" DNA technique police used to hunt the 'Golden State Killer' suspect', *Washington Post*, 27 April. Available at: www.washingtonpost.com/news/true-crime/wp/2018/04/27/golden-state-killer-dna-website-gedmatch-was-used-to-identify-joseph-deangelo-as-suspect-police-say/ (Accessed: 12 May 2021).

St. John, P. (2019) 'DNA genealogical databases are a gold mine for police, but with few rules and little transparency', *Los Angeles Times*, 24 November. Available at: www.latimes.com/california/story/2019-11-24/law-enforcement-dna-crime-cases-privacy (Accessed: 12 May 2021).

St. John, P. (2020) 'The untold story of how the Golden State Killer was found: A covert operation and private DNA', *Los Angeles Times*, 8 December.

Stern, J. & Zhang, S. (2021) 'The victims left behind by genetic genealogy', *The Atlantic*, 27 January. Available at: www.theatlantic.com/science/archive/2021/01/genetic-genealogy-race/616171/.

Toom, V., Granja, R. & Ludwig, A. (2019) 'The Prüm Decisions as an aspirational regime: Reviewing a decade of cross-border exchange and comparison of forensic DNA data', *Forensic Science International: Genetics* (41), pp. 50–57. doi.org/10.1016/j.fsigen.2019.03.023.

University of Minnesota (2020) 'LawSeq: Mapping & shaping the law of genomics', Twin Cities, Minnesota, US. Available at: https://lawseq.umn.edu/ (Accessed: 24 January 2020).

Wong, K. (2019) 'A message to Verogen customers about the GEDmatch partnership', Available at: https://verogen.com/a-message-to-verogen-customers-about-the-gedmatch-partnership/10 December. (Accessed: 24 January 2020).

Zhang, S. (2019) 'The messy consequences of the Golden State Killer case', *The Atlantic*, 1 October. Available at: www.theatlantic.com/science/archive/2019/10/genetic-genealogy-dna-database-criminal-investigations/599005 (Accessed: 12 May 2021).

Epilogue

14

TECHNOLEGAL POLICIES AND PRACTICES

Studying the past, present and future of forensic genetics

David Skinner

The editors of this collection rightly argue that the practice of forensic genetics emerges through the interplay of scientific and legal realms. They suggest the concept of technolegal worlds as a 'methodological tool' that focuses attention on the times, places and spaces where the sociotechnical and the socio-legal converge. Beyond this, when commissioning contributions to this volume, the editors left the definition of *technolegal worlds* open enough to allow authors to articulate their own perspectives. The quality and collective coherence of the chapters in this book confirm that the exploration of the technolegal is worth persevering with. It is my privilege to reflect on the overall lessons of the conjunction of chapters gathered here and highlight directions for further conceptual work.

Truth making across technolegal worlds

A starting assumption of many of the book chapters is the special qualities ascribed to DNA: forensic genetics has achieved unique evidential credibility and invites a level of public interest and institutional oversight that is not shared by other identification technologies. This is a manifestation of a wider genetic exceptionalism that reflects the special qualities ascribed to DNA as the essence of the person. DNA also blurs the distinction between tissue and computerised profile and thus seems to offer resources for analysis that are 'both physically and informationally infinite' (Parry and Gere 2006: 153).

But while genetic data is deemed exceptional in terms of the information it contains and the hopes and the fears it excites, its practical value often seems elusive. Huge claims were made for the potential of science to revolutionise health, society and the economy in the build-up to and at the completion of the mapping of the human genome at the end of the twentieth century. As Jenny Reardon's book *The Postgenomic Condition: Ethics, Justice & Knowledge After the Genome* (2017)

DOI: 10.4324/9780429322358-19

chronicles, however, in the twenty-first century, despite massive investment, ultimate delivery on this promise is seemingly still postponed.

> Rather than reveal meaningful knowledge about life itself, genomics instead has given life to a deluge of data. How to make anything of value out of this data is now, quite literally, the million dollar question.
>
> *(Reardon 2017: 13)*

Forensic genetics has a paradoxical relationship with the wider project to build and exploit genomic knowledge. On the one hand, it is dwarfed by and parasitic on the titanic efforts in biomedicine to extract value from the genome. On the other, it is one of a select number of areas where genomics can claim to make a significant social difference here and now.

The work of forensic genetics, as in other areas of genomics, is not simply about accessing the latent message of DNA. A crucial insight, reiterated in various forms by many of the chapters in this book, is that truth is not inherent in DNA; it is produced within and across scientific and technological and legal worlds. By asking how truth is made, disseminated, and debated in these realms, this collection invites us to revisit the central concerns of key ground-breaking texts of social studies of forensic genetics. These considered how the credibility, validity and reliability of DNA profiling – then a novel technique of identification – were built in the 1980s and 1990s.

Jay Aronson's *Genetic Witness: Science, Law, and Controversy in the Making of DNA* (2007) showed the problems that initially arose when the results of DNA profiling technology transferred from the laboratory to the courtroom and how these were eventually ironed out through critical examination in adversarial criminal trials.

> The development of DNA typing featured scientists weaving together technical claims with legal, social, and political ones, as well as lawyers, politicians, and judges making choices that would seem to require a great deal of scientific knowledge and expertise. The technique itself, standards of 'good science', as well as the relevant expertise needed certify its credibility within the legal system, were significantly altered by this process.
>
> *(Aronson 2007: 4)*

Other work considered the interaction of science and law from a social studies of science and technology perspective. Michael Lynch et al.'s *Truth Machine: The Contentious History of DNA Fingerprinting* (2008) explored how, during the 1980s and 1990s, DNA profiling became the gold standard technique for identification and eventually achieved what they memorably term 'transcendent evidentiary status' (p. 337). While Aronson's book focused primarily on formative 'DNA wars' in the courtroom, Lynch and colleagues' fifteen-year study gave equal weight to developments in the science press and forensic laboratories in the UK and North

America. They showed how alleged sources of uncertainty about DNA evidence 'were bypassed or black boxed by changes in forensic technique, administration, and law' (p. 254) to the extent that, by 1999, profiling trumped human witness and other means of identification.

According to Lynch and colleagues, by the end of the twentieth century, the socio-technology of DNA profiling was 'post-closure' and unlikely to be fundamentally challenged as a method of identification. We should, however, remain curious about the intersection of scientific and legal realms. This is in part because of the importance of other contexts beyond the immediate workings of the laboratory or court that relate to public trust, accountability and operational and ethico-political oversight. A ground-breaking study in this respect was Robin Williams and Paul Johnson's *Genetic Policing: The Uses of DNA in Police Investigations* (2008), which showed how the growth of the UK national police database was underpinned by an 'interwoven series of technical, legislative and organisational changes' (p. 6). Richard Hindmarsh and Barbara Prainsack's 2010 edited collection *Genetic Suspects* further developed this insight by focusing on the importance of national systems of governance to both the public acceptability and legal credibility of forensic genetics.

Aronson's and Lynch and colleagues' assertions about the closure of debate about DNA profiling relate specifically to a first wave of forensic genetics concerned directly with identification via matching genetic material to individual profiles. As the other foundational texts cited previously suggest, however, questions of closure where never as clear cut in relation to a second wave of developments resting on the mass retention of DNA. In the period following their publication, a third wave of innovation relating to the 'prediction' of characteristics of as-yet-unknown suspects prompted further uncertainties and debates (Williams and Wienroth 2014). The chapters in Parts II and III of this collection show how new applications and techniques are objects of what Christopher James Lawless, in his chapter, terms 'contested anticipations' and so invite us to peek under the lid of the black box of DNA profiling once again.

Social studies of forensics continue to be drawn to the novel science and hard cases in law that drive innovation. In-depth studies rooted in ethnography and documentary research, as Roos Hopman et al in this collection put it in their study of the adoption of massive parallel sequencing in forensics, trace the journey of cutting-edge techniques 'from promise to practice'. Although often less attractive to funders, in-depth and in situ research into mundane, established socio-technologies, systems and results at the intersection between law, science and police practice are just as important. Chapters in this collection by Ilpo Helén and Anna-Maria Tapaninen on the use of DNA in migration law and Dana Wilson-Kovacs on the introduction of rapid DNA technologies illustrate the potential of this type of research: showing how uncertainties persist around the application of supposedly entrenched and unassailable techniques to genetic identification.

Despite their emphasis on closure, Lynch and colleagues recognised that 'the exceptional credibility assigned to DNA evidence does not exempt it from the judgements and contingencies that surround its use in criminal investigation' (2008: xvii). Moreover, experience shows that in criminal cases, DNA rarely stands on its own or speaks for itself but is deployed in conjunction with other kinds of evidence (forensic and otherwise) and human testimony. This holds true in other types of legal context where genetic identification has been applied. Olarte-Sierra and Castro Bermúdez in this volume show the complexities of using profiling in post-conflict victim identification in Colombia. In this example, DNA profiling is a last resort and must be joined with other types of evidence in the hope of achieving victim identification and a satisfactory narrative of the circumstances of death. Similarly, Helén and Tapaninen explore the increasingly ambiguous role of DNA profiling in attempts resolve migrant claims for family reunification rights in Finland. In a political atmosphere of growing scepticism towards such claims, DNA seems a panacea for the disbelieved, but evidence of genetic connection is often not enough on its own to demonstrate family belonging.

Truth is produced (and denied) with DNA within and across legal and scientific worlds. Researchers highlight the disjunctions between these realms, particularly issues of communication and translation of knowledge. Just as significant and currently less well explored, however, is the permeability or otherwise of these boundaries and the ways they are performed and maintained. I am struck by Helén and Tapaninen's discussion of the ways in which forensic scientists operate in a 'largely self-referential and self-vindicating world' and detach from the life-shaping consequences of their work for the people whose DNA they analyse. The evidential power of DNA in legal settings arguably rests in part on supposed scientific detachment, but this position can allow scientists (and the researchers who study them) to park important ethical and political questions. Studies of technolegal worlds should consequently include cases in which the boundaries established between experts and lay people are problematised or breached (see, for example, Schwartz-Marin and Cruz-Santiago 2016).

Routine truth making in and across the laboratory and the court is underpinned and facilitated by material, legal and administrative infrastructures; regulation, oversight and governance generate epistemic credibility; and operational confidence. Matthias Wienroth's discussion of the UK case in this volume suggests that what he terms 'systemic trust' is fundamental to the credibility of forensic genetics. He alerts us to the diversity and fragility of trust relationships: DNA gains and maintains value through a combination of symbolic, practical, legislative and formal ethical processes that Wienroth terms 'technologies of trustworthiness'. Police forensic DNA databases, for example, have a practical value that is mutually interdependent with the professional standards of scientists and data custodians; the formal, statutory standing of a database; the admissibility and credibility of the data it holds in law; and legitimacy expressed in wider public support and cooperation. We, therefore, should think about the 'closure' of DNA profiling (and the contingency of that closure) in infrastructural terms: not as a one-off but as something that needs continued maintenance as each of these foundational elements of trust can be challenged.

Genetic citizenship and the worlding of forensic DNA profiling

As their introduction explains, the editors embrace the notion of 'worlds' in part because it allows them to consider the distinctiveness and interconnection of the different realms of forensic genetics in both institutional and geographical terms. Hindmarsh and Prainsack's 2010 volume collected a series of national case studies that, in combination, showed both the remarkable international spread of forensic DNA profiling and the important regional and national variations in the specifics its adoption. A strength of many of the chapters in this book is that, more than a decade on, they invite further reflection on the connections and inequalities between national nodes of now much-expanded networks of international forensic science.

The ubiquity of profiling should not blind us to the importance of the state in the development of forensic genetics as a funder, custodian, user and regulator. Much of the early technolegal research focused on how DNA was mobilised and contested in Anglo American contexts that are adversarial and jury based rather than the judge-led criminal justice systems in many countries, where expert evidence is produced and scrutinised quite differently. Even between ostensibly similar jurisdictions, there are significant differences in the rules and legal frameworks of the harvesting, analyses and storage of DNA samples and profiles for forensic use.

Forensic genetics is, therefore, entangled with the nation as legal entity and as imagined community. DNA databases, for example, are manifestations of the wider phenomenon of genomic nationalism, in which governments support prestige projects that seek value in the curation of the tissue of their populations. (For discussions of this in relation to biomedicine, see Wade et al. 2014; Reardon 2017.) We should reflect on the constructions of bio-citizenship, which inform ethical, legal and social discourse about appropriate collection, retention and use of forensic DNA. These have often unsatisfactorily been framed in terms of the reciprocal rights and obligations in relation to the nation state. The chapter in this volume by Barbara Prainsack and Gabrielle Samuel on policymaking about new techniques of DNA phenotypic profiling is an example of the sort of work urgently needed to broaden the terms of reference, mechanics and participants involved in a democratic discussion of innovation.

Prainsack and Samuel are excited by the potentials of a 'solidarity-infused understanding of security'; other case studies suggest, however, that communitarian and authoritarian impulses can co-exist surprisingly comfortably (Toom 2014). As Prainsack and Samuel acknowledge, generic accounts of citizenship sidestep the ways in which social groups are more or less likely to be the objects of genetic surveillance. A striking example of this relates to race and racism: new techniques in practice reinforce existing patterns of over-policing of ethnic minorities whilst, at the border, migrants are often subject to the unfettered application of new techniques of identification without the protections of the citizen (Skinner 2020). We should be prepared to ask hard questions about the complicity of forensic science in systems that, through their normal workings, produce injustices that have little to do with the accuracy or otherwise of a DNA match.

The national case studies in Hindmarsh and Prainsack's 2010 volume were drawn from Western Europe, the USA and Australasia. This collection includes chapters on Colombia, Brazil, Ghana and South Africa. More needs to be done to unpick the drivers of the spread of DNA profiling beyond its initial sites of development. It has been fostered via international networks of scientists and underpinned by the development of international standards and systems of forensic training. It has also involved the building of infrastructures and institutions designed to support the sharing of data, expertise and knowledge across national criminal justice systems. The role of the technology companies and consulting firms in spreading new policing technologies also needs to be addressed further.

The establishment of forensic laboratories and DNA databases in countries outside 'the West' has become an indicator of national development and funded as such in some cases via Western foreign aid programmes. Such is the credibility of forensic genetics that it is deemed a largely benign or progressive development even in settings where many citizens have good reasons to be cautious about the workings of the police and the courts. This is well illustrated by Noah Tamarkin's account of the embrace of forensic genetics in postapartheid South Africa. He shows how politicians who, because of the recent history of racial oppression, might distrust over-reach by the police, nonetheless enthusiastically adopted DNA profiling, promoted as a means of protecting women and children from abuse and society from the folk devil of the migrant criminal.

The chapters on South Africa, on Ghana by Aaron Opoku Amankwaa and Judith Amankwa Addo and on Brazil by Vitor Simonis Richter and Luiza Louzada are, as Tamarkin puts it, studies of the formation of new technolegal worlds. We cannot that DNA profiling has similar rationales and implications as it travels into settings where elements of the scientific infrastructure and/or criminal justice system are under-developed and over-stressed. We should also be concerned about the take-up of profiling by authoritarian governments. Both Kuwait and Saudi Arabia have floated projects for compulsory whole-population databases. Chinese companies are repurposing medical DNA data for forensic purposes; the Chinese state is collecting the DNA of members of minorities and, in some regions, makes donation of DNA a condition of the issuance of a passport (Wee 2020). These cases are, unfortunately, likely to offer comparator examples of how forensic genetics can operate without independent democratic or judicial control. There are urgent questions about the responsibilities of global science in relation to these developments and the ways in which its techniques, institutions and notions of neutrality as currently formulated may facilitate injustice.

Unravelling technolegal worlds?

The chapters in this volume are testimony to the richness of the social studies of forensic genetics. Academics in this now well-established field have been active participants in debates that have accompanied the ongoing development and dissemination of the socio-technology of profiling. Some have become interlocutors and experts who help define the parameters of innovation and the legislative and ethical systems of governance around forensic genetics. Changes that straddle the technological and the legal are afoot, however; these challenge some of the underlying assumptions of the field.

One of these assumptions is the genetic exceptionalism discussed at the start of this piece. DNA profiling is deemed (for good or ill) a uniquely powerful means of identification able to deliver a step-change in criminal justice. We can return to this claim with the benefit of now considerable hindsight. Carole McCartney and Aaron Opoku Amankwaa's examination of the 'effectiveness' of the UK national police DNA database in this volume is a rare critical examination of the practical impact of supposed game-changing genomics. These authors argue that the value and optimal size of DNA databases remain hard to quantify as genetic analysis is directly linked to the solution of a relatively small number of cases. As this discussion shows, cost-benefit analyses of investment in forensic genetics depend on expanding criteria for its success to more nebulous and politicised outcomes of deterrence and public security.

McCartney and Amankwaa rightly point out that supporters find it easiest to celebrate the effectiveness of forensics genetics via narratives of hard cases solved rather than quantification of justice delivered. In this respect, we can also legitimately ask the relevance of cutting-edge genetics to the major challenges in criminal justice. There is a particular irony to the ways in which the significance of profiling and limits on the ethical acceptability of its application are frequently discussed in relation to the solving of cases of the rape and killing of women by strangers. Globally, the phenomenon of violence against women remains chronically under-addressed. Taking the UK example, in 2019, official data reported that annually, 1.6 million women in England and Wales had experienced domestic abuse and that, on average, two women a week are killed by a current or former partner (ONS 2020). The England and Wales Crown Prosecution Service reported that the number of people receiving a custodial sentence after conviction for rape declined from 1,058 in 2010 to 537 in 2020.

The example of sexual assault and violence against women points to challenges which cannot meaningfully be addressed through better science. The failings discussed here also have a common context: a criminal justice system struggling to cope with the changing demands placed on it. In the UK, this sense of crisis extends to its previously world-leading forensic science provision that is now suffering from falling standards, under-capacity and under-funding (House of Lords Science and Technology Select Committee 2019).

The uniqueness of DNA analysis is often said to rest on its capacity to discover meaningful evidence where none was previously thought to exist. Many emerging challenges in policing relate to a contrasting problem of how to manage an over-abundance of suspects and evidence. The ready availability of digital images and traces, in particular, generates a near-limitless pools of potentially relevant information which require new forms of management and risk assessment. Innovation in forensic genetics, such as the use of genealogy websites discussed by Erin Murphy in her contribution to this volume, is likely to take place within a wider array of responses to this issue of data deluge. This repurposing of DNA records and other developments, including DNA phenotyping, the use of partial 'familial' DNA matches and new techniques for analysis of mixed samples, is an attempt to generate potentially useful investigatory leads from imperfect genetic source material, gathered from a variety of sources. These developments together raise new questions about their accuracy, the control of data and the speculative construction of suspect populations. There is a reconfiguration of technological worlds on the horizon.

References

Aronson, J. (2007) *Genetic Witness: Science, Law, and Controversy in the Making of DNA*. Rutgers University Press.

Hindmarsh, R. and Prainsack, B. (eds) (2010) *Genetic Suspects: Global Governance of Forensic DNA Profiling and Databasing*, Cambridge University Press.

House of Lords Science and Technology Select Committee (2019) *Forensic Science and the Criminal Justice System: A Blueprint for Change*, https://publications.parliament.uk/pa/ld201719/ldselect/ldsctech/333/33302.htm.

Lynch, M., Cole, S., McNally, R. and Jordan, K. (2008) *Truth Machine: The Contentious History of DNA Fingerprinting*, The University of Chicago Press.

Office of National Statistics (2020) www.ons.gov.uk/peoplepopulationandcommunity/crimeandjustice/bulletins/domesticabuseinenglandandwalesoverview/.

Parry, B. and Gere, C (2006) 'Contested bodies: Property models and the commodification of human biological artefacts', *Science as Culture*, 15(2), pp. 139–158.

Reardon, J. (2017) *The Postgenomic Condition: Ethics, Justice & Knowledge After the Genome*, University of Chicago Press.

Schwartz-Marin, E. and Cruz-Santiago, A. (2016) 'Pure corpses, dangerous citizens: Transgressing the boundaries between experts and mourners in the search for the disappeared in Mexico', *Social Research*, 83(2), pp. 483–510.

Skinner, D. (2020) 'Race, racism and identification in the era of technosecurity', *Science as Culture*, 29(1), pp. 77–99.

Toom, V. (2014) 'Trumping communitarianism: Crime control and forensic DNA typing and databasing in Singapore', *East Asian Science, Technology and Society: An International Journal*, 8(3), pp. 273–296.

Wade, P., López-Beltrán, C., Restrepo, E. and Ventura-Santos, R. (2014) *Mestizo Genomics*, Duke University Press.

Wee, S.-L. (2020) 'China is collecting DNA from tens of millions of men and boys, using US Equipment', *The New York Times*, www.nytimes.com/2020/06/17/world/asia/China-DNA-surveillance.html.

Williams, R. and Johnson, P. (2008) *Genetic Policing: The Uses of DNA in Police Investigations*, Willan.

Williams, R. and Wienroth, M. (2014) *Ethical, Social and Policy Aspects of Forensic Genetics: A Systematic Review*, EUROFORGEN.

INDEX

Note: Numbers in **bold** indicate a table

For Product Safety Concerns and Information please contact our EU
representative GPSR@taylorandfrancis.com
Taylor & Francis Verlag GmbH, Kaufingerstraße 24, 80331 München, Germany

www.ingramcontent.com/pod-product-compliance
Ingram Content Group UK Ltd.
Pitfield, Milton Keynes, MK11 3LW, UK
UKHW020143200926
475702UK00005B/883
9780367338497